Herbert West Seager

Natural History
in
Shakespeare's Time
being
Extracts illustrative of the Subject
as he knew it

Elibron Classics
www.elibron.com

Elibron Classics series.

ISBN 0-543-95951-1 (paperback)
ISBN 0-543-95950-3 (hardcover)

This Elibron Classics Replica Edition is an unabridged facsimile of the edition published in 1896 by Elliot Stock, London.

NATURAL HISTORY
IN SHAKESPEARE'S TIME.

ARISE PRAY WORK

Natural History

in

Shakespeare's Time:

being

Extracts illustrative of the Subject
as he knew it.

Made by

H. W. Seager, M.B., &c.

Also Pictures thereunto belonging.

"Lege, disce."

Printed for

Elliot Stock,

and are to be sold at 62, Paternoster Row,
LONDON.

1896.

PREFACE.

THIS book presents in a convenient form for reference a collection of the quaint theories about Natural History accepted by Shakespeare and his contemporaries. The work is meant to be rather a sketch than an exhaustive treatise, otherwise it would fill many volumes. The plan of the book is to give some illustration of each word mentioned by Shakespeare when there is anything remarkable to be noted about it. The term "Natural History" has been taken in its widest sense, as including not only fauna but flora, as well as some precious stones.

It is certain that Shakespeare believed some of the strange ideas here mentioned, especially about those animals which he had had no opportunity of observing in their wild state; but, on the other hand, Shakespeare's knowledge of Natural History (in so far as his own observation extended) was far greater than that of his contemporaries as here illustrated.

All the quotations inserted in this book are from works which were the standard authorities in Shakespeare's time, and the extracts are cited with the utmost exactness, except where the spelling in all but a few rare words has been modernized, and where uninteresting matter has been omitted. A few of these extracts are given, not for their contents, but for their style. Here and there illustrative notes have been added.

The text of Shakespeare referred to is that of the "Globe" edition.

The books and editions most frequently quoted are:

Bartholomew. "Liber de proprietatibus rerum editus a fratre Bartholomeo anglico ordinis fratrum minorum. Impressus Argentine Anno domini MCCCCLXXXV. Finitus in die Sancti Valentini."

Bartholomew (Berthelet). "Bartholomeus de Proprietatibus Rerum." [Translated into English by J. Trevisa.] "In aedibus T. Bertheletti, Lond. 1535." So runs the description in the British Museum catalogue, but this version does not follow Trevisa's translation accurately; on the contrary, it quotes Trevisa for some deviations from, and additions to, Bartholomew's text.

Batman. "Batman upon Bartholome, his Booke 'De Proprietatibus Rerum'" [in the translation by J. Trevisa]. "Newly corrected, enlarged and amended; with such Additions as are requisite vnto euery seuerall Booke. Taken foorth of the most approued Authors, the like heretofore not translated in English," etc. T. East, Lond. 1582. Probably Shakespeare used not Batman's version, but the Berthelet edition, which, being older, would probably be cheaper in his days. All of Batman's "Additions" that are of any interest are quoted in these pages, but they are few and generally unimportant. His emendations consist mostly in the substitution for an archaic word of a more modern and less interesting one.

There can be no doubt that Friar Bartholomew's book was the standard authority on Natural History in Shakespeare's youth; indeed, it was the only popular authority. It is true that there were some few books on Natural

History in Shakespeare's time, which were written on more scientific principles, *e.g.*, "Carol. Clusii Exoticorum Libri Decem," but these were published abroad, and in Latin, and were probably unknown to him.

Hortus Sanitatis. The edition here quoted is one of the two undated ones mentioned by Hain, and from it the woodcuts have been photographed. There were five dated editions between 1490 and 1517.

Topsell. "The History of Four-footed Beasts and Serpents and [*T. Mouffet*] The Theater of Insects"; in one volume, Lond. 1658. The first edition of Topsell's "History of Four-footed Beasts" was printed in 1607. The "Serpents" followed in 1608. Mouffet's "Theater of Insects" appears to have been written about 1584.

Holland's Pliny. "The Historie of the World, commonly called the Naturall Historie of C. Plinius Secundus. Translated into English by Philemon Holland, Doctor of Physicke . . . London, 1634." The first edition of this book was published in 1601.

Harrison's "Description of Britain," sometimes quoted as *Holinshed's* "Description of Britain." "An Historicall description of the Iland of Britaine . . . written by W. H." [William Harrison]. This description is prefixed to the first volume of Holinshed's Chronicles, and is here quoted from the 1586 edition. It first appeared in 1577.

Gerard's "Herbal." "The Herball or Generall Historie of Plantes. Gathered by John Gerarde of London, Master in Chirurgerie. Very much Enlarged and Amended by Thomas Johnson, Citizen and Apothecarye of London. London, 1633." The first edition of Gerard's "Herbal" appeared in 1596. [Johnson's additions are in this book generally distinguished by his name.]

Parkinson's "Herbal." "Theatrum Botanicum: The Theater of Plants. Or an Herball of a large Extent . . .

by John Parkinson, Apothecary of London, and the King's Herbarist. London, 1640."

Albertus Magnus. "Alberti cognomento Magni . . . De Virtutibus Herbarum, Lapidum et Animalium quorundam, libellus. Item, De Mirabilibus Mundi, ac de quibusdam effectibus causatis a quibusdam animalibus," etc. Lugduni, 1553.

Lupton. "A thousand Notable things of sundry sortes: Whereof some are wonderfull, some strange, some pleasant, divers necessary, a great sort profitable, and many very precious." London, 1627. The first edition of this book was published in 1595, the second in 1601.

Hart. "ΚΛΙΝΙΚΗ, or the Diet of the Diseased . . . by James Hart, Doctor in Physicke." London, 1633.

Evelyn. "Silva, or a Discourse of Forest-trees, . . . Also Acetaria; or a Discourse of Sallets, with Kalendarium Hortense; or The Gard'ner's Almanack . . . by John Evelyn, Esq.," etc. Fourth edition. London, 1706.

The edition of Ben Jonson here quoted is the folio of 1692; that of Beaumont and Fletcher the folio of 1679. The other plays cited are from modern reprints.

I have to thank Mr. Mihill Slaughter for the care with which he has photographed the woodcuts in the *Hortus Sanitatis*.

HERBERT W. SEAGER.

HAMPTON COURT,
December, 1896.

SHAKESPEARE'S NATURAL HISTORY.

Aconitum.

> Shall never leak though it do work as strong
> As aconitum or rash gunpowder.
> ii. KING HENRY IV., iv. 4, 47-8.

GERARD, in his "Herbal," says that the poison of the broad-leafed and mountain wolf's-bane "is of such force that, if a man especially, and then next any four-footed beast, be wounded with an arrow or other instrument dipped in the juice hereof, they die within half an hour after, remediless"; but the winter wolf's-bane "is not without his peculiar virtues. It is reported to prevail mightily against the bitings of scorpions, and is of such force that, if the scorpion pass by where it groweth, and touch the same, presently he becometh dull, heavy and senseless; and if the same scorpion by chance touch the white hellebore, he is presently delivered from his drowsiness." He enumerates in all twelve varieties of Aconitum, or wolf's-bane, and in addition "mithridate," or wholesome wolf's-bane (*Anthora*) which is the Bezoar, or counter-poison to Aconite. Gerard says further that, according to Avicenna, "the mouse nourished and fed up with *Napellus* (*Monk's-hood*) is altogether an enemy to the poisonsome nature thereof, and delivereth him that hath taken it from all peril and danger." But Antonius Guanerius of Pavia "is of opinion that it is not a mouse that Avicen speaks of, but a fly," which is found on the leaves of wolf's-bane, and from which an antidote is to be made with bay-berries, mithridate, honey, and oil of olive.

[Honey is put in opposition to "mortal Aconite" by *Dekker* in "The Devil's Answer to Pierce Pennylesse."

Aconitum originated from the foam of Cerberus. *V. Heywood's* "Brazen Age."]

Adamant.

As true as steel, as plantage to the moon,
As sun to day, as turtle to her mate,
As iron to adamant, as earth to the centre.
TROILUS AND CRESSIDA, iii. 2, 184-6.

You draw me, you hard-hearted adamant;
But yet you draw not iron, for my heart
Is true as steel; leave you your power to draw,
And I shall have no power to follow you.
MIDSUMMER NIGHT'S DREAM, ii. 1, 195-9.

V. **Diamond.**

ADAMAS is a little stone of Ind, and is coloured as it were iron, and shineth as crystal; but it passeth never the quantity of a walnut. No thing overcometh it, neither iron nor fire. And also it heateth never. But though it may not be overcome, and though it despise fire and iron, yet it is broke with new hot blood [of a he-goat (*Bartholomew*)]. This stone is contrary to Magnes. For if an Adamas be set by iron, it suffereth not the iron come to the Magnes, but it draweth it by a manner of violence from the Magnes, so that though the Magnes draweth iron to itself, the Adamas draweth it away from the Magnes. It is said that this stone warneth of venom as Electrum doth; and putteth off divers dreads and fears, and withstandeth witchcraft. Dioscorides saith that it is called a precious stone of reconciliation and of love. For if a woman be away from her husband, or trespasseth against him: by virtue of this stone she is the sooner reconciled to have grace of her husband. And hereto he saith, that if a very Adamas be privily laid under a woman's head that sleepeth: her husband may wit whether that she be chaste or no. For if she be chaste by virtue of that stone she is compelled in her sleep to beclip [embrace] her husband; and if she be untrue, she leapeth from him out of the bed, as one that is unworthy to abide the presence of that stone. Also, as Dioscorides saith, the virtue of such a stone borne in the left shoulder, or in the left arm-pit, helpeth against enemies, against woodness, chiding, and strife, and against fiends that noy [annoy] men that dream in their sleep, against fantasy, against swevens [dreams] and venom.

Bartholomew (*Berthelet*), bk. xvi. § 9.

THERE is nowadays a kind of Adamant which draweth unto it flesh, and the same so strongly, that it hath power to knit and tie together two mouths of contrary persons, and draw the heart of a man out of his body without offending any part of him.

Edward Fenton's "Certaine Secrete Wonders of Nature" (apud Steevens).

Of the Magnet Bartholomew says:

MAGNES is a stone of Ind, coloured somewhat as iron. And is found in Ind among the Troglodytes, and draweth to itself iron in such wise, that it maketh as it were a chain of iron rings. Also it is said, that it draweth glass molten as it doth iron. In certain temples is made an image of iron, and it seemeth that that image hangeth in the air.

And in Ethiopia is another kind of Magnes that forsaketh iron, and driveth it away from him. Also the same Magnes draweth iron to it in one corner, and putteth it away in another corner. And the more blue the Magnes is the better it is.

[He then ascribes to it the same virtues as belong to the Adamant—of reconciling men and their wives, and testing women's chastity.]

If the powder thereof be sprung and done upon coals in four corners of the house, it shall seem to them that be in the house, that the house should fall anon. And that seeming is by moving that cometh by turning of the brain. And there be mountains of such stones, and therefore they draw to them and break ships that be nailed with iron [of which Sir John Mandeville also speaks].

Bartholomew (*Berthelet*), bk. xvi. § 63.

[It is evident from these quotations that Shakespeare and Lylly confused the Adamant or Diamond, which was supposed to repel iron, with the iron-attracting Magnet, being no doubt misled by the similarity of their other properties.]

If this stone be placed on coals in the four corners of the house, I say, if it be pounded and sprinkled on the coals, sleepers will flee the house and quite forsake it, and then thieves can see after all that they please.

Albertus Magnus, "Of the Virtues of Stones."

Adder.

Is the adder better than the eel,
Because his painted skin contents the eye?

TAMING OF THE SHREW, iv. 3, 179, 180.

Art thou like the adder waxen deaf?

ii. KING HENRY VI., iii. 2, 76.

It is the bright day that brings forth the adder;
And that craves wary walking.

JULIUS CÆSAR, ii. 1, 14, 15.

Each jealous of the other, as the stung
Are of the adder.

KING LEAR, v. 1, 56, 57.

An Adder dwelleth in shadows, he slideth and wriggleth in slipper draughts and wrinkles, and in slimy passing.

The Adder fleeth the hind, and slayeth the lion, and he eateth rue, and changeth his skin, and loveth hollowness of wood and of trees, and drinketh milk busily. And he hurteth and grieveth with the teeth, and with the tail, and sheddeth venom, and lieth in the sun under hedges, and sucketh bitches, eateth flies, and licketh powder [dust]. The grease of the water-adder helpeth against the biting of the crocodile; and if a man have with him the gall of this adder, the crocodile shall not grieve him nor noy him; and that most jeopardous and fearful beast dare not, nor may do against him in no manner of wise damage nor grief, which beareth the gall of the said Adder.

Bartholomew (*Berthelet*), bk. xviii. § 34.

V. Aspick, Serpent.

Agate.

If low, an agate very vilely cut.
MUCH ADO ABOUT NOTHING, iii. 1, 65.

I was never manned with an agate till now; but I will inset you neither in gold nor silver, but in vile apparel, and send you back again to your master for a jewel.

ii. KING HENRY IV., i. 2, 18-23.

Agate-ring.
i. KING HENRY IV., ii. 4, 78.

In shape no bigger than an agate-stone
On the forefinger of an alderman.
ROMEO AND JULIET, i. 4, 55-6.

THE first manner thereof helpeth witchcraft. For therewith tempest is changed; and stinteth rivers and streams. And the manner kind of Creta changeth perils and maketh gracious and pleasing, and fair showing and speaking, and giveth might and strength. The third manner stone, that is of Ind, comforteth the sight, and helpeth against thirst and venom, and smelleth sweet if it be nigh. The burning of it is odoriferous. *Bartholomew* (*Berthelet*), bk. xvi. § 11.

[Agates were worn by justices of the peace.] Thou wilt spit as formally, and show thy Agate and hatched chain, as well as the best of them.

Beaumont and Fletcher's "Coxcomb". (Steevens).

[The Agate which is found in the eagle's nest is of two sorts, male and female.] The male thereof is hard, and is somewhat blazing. And the female is nesh [soft]. Also this stone containeth and breedeth another stone within him. The virtue of this stone maketh a man sober, and augmenteth and increaseth riches, and so it doth love, and helpeth greatly to obtain and conquer victory and favour. If there be any man suspect of fraud of poisoning, if he be guilty, this stone put under his meat will not suffer him to swallow his meat, and if the stone be withdrawn, he shall not tarry to swallow his meat.

Bartholomew (Berthelet), bk. xvi. § 39.

Alabaster.

Why should a man, whose blood is warm within,
Sit like his grandsire cut in alabaster?

MERCHANT OF VENICE, i. 1, 83-4.

THIS stone helpeth to win victory and mastery. This gendereth and keepeth friendship.

Bartholomew (Berthelet), bk. xvi. § 3.

Almond.

The parrot will not do more for an almond.

TROILUS AND CRESSIDA, v. 2, 193.

["Almond for parrot" is a proverbial phrase so common as to need no reference.

Almond milk was made of Almonds with neck of mutton, barley, herbs, and salt ("The Good House Wives Treasurie").

Almond butter was eaten in Lent, and also used as a cosmetic for the hands (*Ben Jonson's* "Staple of News," and *Shirley's* (?) "Andromana").

Paste of Almonds is also mentioned in the "Staple of News."]

Amber. [LAT. *Electrum.*]

Her amber hair for foul hath amber quoted.

LOVE'S LABOUR'S LOST, iv. 3, 87.

With amber bracelets, beads, and all this knavery.

TAMING OF THE SHREW, iv. 3, 58.

ELECTRUM is a metal, and is more noble than other metals. And hereof be three manner of kinds—one is such, that when it runneth first out of the tree, it is fleeting and thin gum, but afterwards with heat or with cold it is made hard as a clear stone, as it were crystal.

That other manner kind is called metal, and is found in the earth, and is had in price. The third manner is made of the three parts of gold, and of the fourth of silver. And kind Electrum warneth of venom, for if one dip it therein, it maketh a great chirking noise, and changeth oft into divers colours as the rainbow, and that suddenly.

Bartholomew (*Berthelet*), bk. xvi. § 38.

THE Amber that is brought from these parts [Konigsberg and Kurland] lies in great quantity scattered on the sand of the sea, yet it is as safe as if it were in warehouses, since it is death to take away the least piece thereof. At Dantzic I did see two polished pieces thereof, which were esteemed at a great price, one including a frog with each part clearly to be seen, (for which the King of Poland then being there offered five hundred dollars) the other including a newt, but not so transparent as the former.

Fynes Moryson, "Itinerary," pt. iii. bk. ii. ch. 3, p. 81.

ANY kind of Amber being sodden in the grease of a sow that gives suck to young pigs, is not only thereby the clearer but also much the better. *Lupton*, bk. i. § 25.

OUR drink shall be prepared gold and Amber.

Ben Jonson's "Fox," iii. 7.

HE must drink his wine
With three parts water, and have Amber in that too.

Ben Jonson's "Magnetic Lady," iii. 2.

[Ambergris was a synonym for Amber, and was also used in caudles, cullises, and comfits.]

I WONDER most at Sophocles the tragical poet. For he sticketh not to avouch, That beyond India Amber proceedeth from the tears that fall from the eyes of the birds.

Holland's Pliny, bk. xxxvii. ch. 2.

AMBER is found as well in other places as in India. Garcias thinks it to be the nature of the soil, as chalk, bole-ammoniac, etc., and not the seed of the whale, or issuing from some fountain in the sea.

Purchas' "Pilgrims," p. 508 (ed. 1616).

Anchovies.

Item, Anchovies and sack after supper. 2s. 6d.
i. King Henry IV., ii. 4, 588-9.

Anchovies, 6 sh. I swear but a saucer full.
Brome, "The Covent Garden Weeded."

He doth learn to make strange sauces, to eat Anchovies, Maccaroni, Bovoli, Fagioli, and Caviare, because he loves 'em. *Ben Jonson*, "Cynthia's Revels," ii. 3.

Clem. For twelve pennyworth of Anchovies, eighteen-pence.

Bess. How can that be?

Clem. Marry, very well, mistress; twelvepence Anchovies, and sixpence oil and vinegar. Nay, they shall have a saucy reckoning. *Heywood's* "Fair Maid of the West," ii. 2.

In midst of meat they present me with some sharp sauce or a dish of delicate Anchovies, or a caviare.
"Lingua," ii. 1.

He feeds now upon sack and Anchovies.
G. Wilkins, "Miseries of Enforced Marriage," iii.

Anchovy (Gr. ἐγκρασίκολος) is a fish nearly like a sardine, so called because it has bile in its head—from ἐν (in), κέρας (head), and χόλος (bile). *Minsheu's* Dictionary, *s.v.*

Animal.

All that is comprehended of flesh and of spirit of life and so of body and soul is called animal—a beast—whether it be airy as fowls that fly, or watery as fish that swim, or earthy as beasts that go on the ground and in fields, as men and beasts, wild and tame, or other that creep and glide on the ground. Some beasts have blood and some have none, as bees and all other beasts with rivelled bodies. But such beasts have other humour in stead of blood.

It is said that in Ind is a beast wonderly shape[d], and is like to the bear in body and in hair and to a man in face. And hath a right red head, and a full great mouth, and an horrible, and in either jaw three rows of teeth distinct between [*i.e.*, separate]. The outer limbs thereof be as it were the outer limbs of a lion, and his tail is like to a wild scorpion with a sting, and smiteth with hard bristle-pricks as a wild swine, and hath an horrible voice as the voice of a trump, and he runneth full swiftly, and eateth men. And among all beasts of the earth is none found more cruel nor more wonderly shapen.

The dolphin and other manner of fish fall to the bottom suddenly, as it were in epilepsy, when they hear sudden thundering, or great moving and noise, and be taken as they were drunk. And fish fleeth and voideth the place of washing and slaughter of other fish, and the blood of other fish, and flee and void also hoary and unclean nets; and come gladly into new.

The female bear bringeth forth a lump of flesh not divided by shape of members; and she keepeth that lump hot under her arm-pits as the hen sitteth on her eggs. And the female bear licketh that lump of flesh and shapeth it some and some, until it receive perfect figure and shape of a bear. Also the panther and the lioness bringeth forth whelps but not complete nor perfectly shapen. In all beasts that bring children forth uncomplete and unperfect, the cause is gluttony, for if kind would abide unto they were complete and perfect, the children would slay the mother with sucking, for immoderate and over-passing appetite.

Fish in one month waxeth fat, and soon afterward wax lean. And some waxeth fat in the northern wind, as fish with long bodies, and some in southern wind, as fish with broad bodies, and some in rain-time. Rain-water accordeth to all manner shell-fish, out-take[n] the fish that hight Roitera [or Koytea—an unidentified class of fish], that dieth in the same day, if he taste rain-water. And too much rain-water grieveth some fish, for it blindeth them.

Some beasts be ordained for man's mirth, as apes and marmosets and popinjays; and some be made for exercitation of man, for man should know his own infirmities and the might of God. And therefore be made flies and lice; and lions and tigers and bears be made that man may by the first know his own infirmity, and be afeard of the second. Also some beasts be made to relieve and help the need of many manner infirmities of mankind—as the flesh of the adder to make treacle. Wolves flee from him that is anointed with lion's dirt [kidney-fat]. If the tail of an old wolf be hanged at the cows' stall the wolves will not come there nigh. Bear's eyes taken out of the head, and bound together under the right arm of man, abateth the fever quartan. Also the long teeth of a wolf healeth lunatic men. Tame four-footed beasts dread and flee if they see a wolf's eye taken out of the head. If thou besmokest the house with the lungs of an ass thou cleansest the house of serpents and other creeping worms.

Bartholomew (*Berthelet*), bk. xviii. § 1.

Ant.

Sometimes he angers me
With telling me of the moldwarp and the ant.

i. KING HENRY IV., iii. 1, 149.

We'll set thee to school to an ant, to teach thee there's no labouring i' the winter.

KING LEAR, ii. 4, 68.

SLEIGHT and business of them is much. For in summer they gather store by the which they may live in winter;

and they gather wheat and reck not of barley, and when the wheat is berained, that they gather to heap, then the Ants do all the wheat out into the sun, that it may be dried again. And it is said that in Ethiopia be Ants shap[ed] as hounds, and diggeth up golden gravel with their feet, and keep it that it be not taken away. And pursueth anon to the death them that take it away. And when they be overset in their houses to be taken, then shed they venomous water upon men, and that water burneth his hand that it toucheth, and breedeth therein itching and

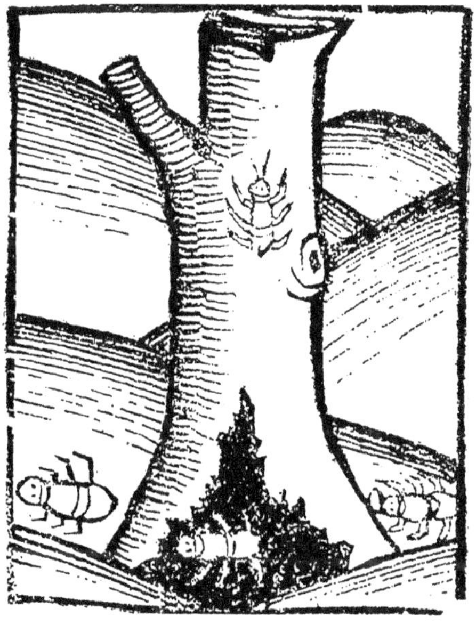

smarting. For they have that water instead of weapon and of armour. In Ind be right great Ants with horns, that keep gold and precious stones with wonder covetise and desire, but Indians steal them in summer-time when the Ants be hid in hills for strong burning heat; but the Ants fly after them busily, which take away the gold; and wound them after, though they flee the Ants riding on swift camels—in them is so wicked fierceness for lust of gold. When bears be sick, they seek Ants, and devour them, and heal themselves in that wise. But in some case Ants' eggs be medicinable. *Bartholomew* (*Berthelet*), bk. xviii. § 53.

[After the account given above of the way in which Indians get the gold from the ants, Sir John Mandeville adds]:

AND in other times when it is not so hot, and that the Pismires [Ants] ne rest them not in the earth, then they get gold by their subtilty. They take mares that have young colts or foals, and lay upon the mares void vessels made therefor; and they be all open above, and hanging low to the earth; and then they send forth the mares for to pasture about those hills, and withhold the foals with them at home. And when the Pismires see those vessels they leap in anon, and they have this kind, that they let nothing be empty among them, but anon they fill it, be it what manner of thing that it be; and so they fill those vessels with gold. And when that the folk suppose that the vessels be full they put forth anon the young foals, and make them to neigh after their dams; and then anon the mares return towards their foals, with their charges of gold; and then men discharge them, and get gold enough by this subtility. For the Pismires will suffer beasts to go and pasture amongst them; but no man in no wise (chap. xxx.).

If you stamp Lupins (which are to be had at the Apothecaries') and therewith rub round about the bottom or lower part of any tree, no Ants or Pismires will go up and touch the same tree.

Lupton's "Notable Things," bk. iv. § 77.

IF you burn the shells of snails with Styrax, and then sprinkle thereof upon an Ants'-hill, thereby they will be driven forth of the ground or place where they are.

Ibid., bk. x. § 77.

V. Pismire.

Ape.

MIDSUMMER NIGHT'S DREAM, ii. 2, 181.

CYMBELINE, i. 6, 39.

APES have knowledge of elements, and be sorry in the full of the moon, and be merry and glad in the new of the moon. Of Apes be five manner kinds, of whom some have tails; and some be like to an hound in the face, and in the body like to an Ape. Some be rough and hairy, and forgetteth soon wildness. And some be pleasing in face with merry movings and playings, and resteth but little. And some be unlike to that other nigh in all manner points, for in the face is a long beard, and have a broad tail. That kind of Apes is next to man's shape, and be diverse and distinguished by tails, and labour wonderly and busily to do all thing that they see: and so oft they shoe themselves with shoes that hunters leave in certain places slyly, and be so taken the sooner; for while they would fasten the thong of the shoe, and would put the shoes on their feet, as they see the hunters do, they be oft taken with hunters ere they may unlace the shoes, and be delivered of them. The Ape is tamed and chastised by violence with beating and with chains, and is refrained with a clog, so that he may not run about freely at his own will, to abate his fierceness and outrage. And the Ape eateth all manner of meats and unclean things, and therefore he seeketh and looketh worms in men's heads, and throweth them into his mouth, and eateth them. The lion loveth Ape's flesh, for by eating thereof he recovereth when he is sore sick.

Bartholomew (*Berthelet*), bk. xviii. § 96.

The Ape ever killeth that young one which he loveth most with embracing it too fervently.

Greene's "Thieves Falling Out," etc.

I'll teach you
To come aloft and do tricks like an ape.

[V. *Massinger*, "The Bondman," iii. 3, for various tricks taught to the ape.]

[Katharina, in "The Taming of the Shrew" (ii. 1, 34)—

I must dance barefoot on her wedding-day,
And for your love to her lead Apes in hell—

alludes to the old proverb:

Such as die maids do all lead Apes in hell—

Compare Douce's note on this passage.]

If you wish to frighten any man while asleep, put the skin of an Ape under his head.

Albertus Magnus, "Of the Wonders of the World."

THE pepper-trees are great, and abound with Apes, who gather the pepper for the Indians gratis, brought thereunto by a wile of the Indians, who first gather some, and lay it on heaps, and then go away, at their return finding many the like heaps made by the emulous Apes.

Purchas' "Pilgrims," p. 457 (ed. 1616).

Apple.

Not yet old enough for a man, nor young enough for a boy; as a squash is before 'tis a peascod, or a codling when 'tis almost an apple.

TWELFTH NIGHT, i. 5, 165-7.

She's as like this as a crab's like an apple.

KING LEAR, i. 5, 15-6.

[Gerard engraves the following sorts of apples: The Pomewater, the Baker's Ditch, the Queening or Queen of Apples, the Summer Pearmain, the Winter Pearmain.

Shakespeare mentions or alludes to several sorts of apples, viz., Apple-john, Pomewater, Codling, Carraway, Leather-coat, Lording, Pippin, Bitter-sweet, and Crab (*q.v.*).]

Apple-john.

I am withered like an old apple-john.

i. KING HENRY IV., iii. 3, 4-5.

The prince once set a dish of apple-johns before him, and told him there were five more Sir Johns, and pulling off his hat, said: 'I will now take my leave of these six, dry, round, old, withered knights.'

ii. KING HENRY IV., ii. 4, 4-9.

[In Heywood's "Fair Maid of the Exchange," Fiddle the clown takes it in snuff when he is called "russeting" and "apple-john."]

THIS apple will keep two years, but becomes very wrinkled and shrivelled.

Steevens' note, ii. KING HENRY IV., ii. 4, 4.

Apricock.

Feed him with apricocks and dewberries.

MIDSUMMER NIGHT'S DREAM, iii. 1, 169.

Go, bind thou up yon dangling apricocks.

KING RICHARD II., iii. 4, 29.

[In 1633 five sorts of Apricots were known: "The common, the long and great, the musk, the Barbary, and the early Apricock."] *Johnson's* edition of *Gerard's* "Herbal," p. 1448.

Ash.

That body, where against
My grained ash an hundred times hath broke.
CORIOLANUS, iv. 5, 112.

ASH is good for shafts and spears. The leaves thereof helpeth against venom, and the juice thereof wrung and drunk helpeth best against serpents. And Ash hath so great virtue, that serpents come not in the shadow thereof in the morning nor at even. And if a serpent be set between a fire and Ash-leaves, he will flee into the fire sooner than into the leaves. In Greece the leaves thereof is poison to beasts, and grieveth not other beasts that chew their cud, and grieveth not beasts in Italy.

Bartholomew (*Berthelet*), bk. xvii. § 62.

THE fruit like unto cods is termed in English Ash-keys, and of some Kite-keys. It is a wonderful courtesy in nature that the Ash should flower before these serpents appear, and not cast his leaves before they be gone again. Three or four leaves of the Ash-tree taken in wine each morning from time to time do make those lean that are fat, and keepeth them from feeding which do begin to wax fat. *Gerard's* "Herbal," *s.v.*

(WHETHER by the power of magic or nature I determine not) I have heard it affirmed with great confidence, and upon experience, that the rupture to which many children are obnoxious, is healed by passing the infant through a wide cleft made in the bole or stem of a growing Ash-tree, through which the child is made to pass; and then carried a second time round the Ash, caused to repass the same aperture again, that the cleft of the tree suffered to close and coalesce, as it will, the rupture of the child, being carefully bound up, will not only abate, but be perfectly cured. The white and rotten dotard part composes a ground for our gallants' sweet powder.

Evelyn's "Sylva," p. 62 (ed. 1706).

Aspick (*i.e.*, Asp).

The pretty worm of Nilus there
That kills and pains not.

* * * * *

Have I the aspic in my lips? Dost fall?
If thou and nature can so gently part,
The stroke of death is as a lover's pinch,
Which hurts, and is desired.

* * * * *

This is an aspic's trail; and these fig-leaves
Have slime upon them, such as the aspic leaves
Upon the caves of Nile.

ANTONY AND CLEOPATRA, V 2, 243-4, 296-9, 354-7.

V. Adder, Serpent.

ASPIS is an Adder [*q.v.*] worst and most wicked in venom and in biting; he casteth out fleeing venom,

and spitteth and springeth out venom by bitings. Of Adders that hight Aspis be divers manner kind, and have diverse effects and doings to noy and to grieve, that is to wit, Dipsas,—when he biteth, he slayeth with thirst. Ipalis is a manner Adder that slayeth with sleep. These manner Adders Cleopatra laid by her, and passed out of the life by death as it were by sleep. And there be many other Adders, and the venom of them is so strong, that they slay with their venom him that toucheth them with a spear. The Adder Aspis, when she is charmed by the enchanter, to come out of her den by charms and conjurations, for she hath no will to come out, layeth her one ear to the ground, and stoppeth that other with her tail, and so she heareth not the voice of the charming, nor cometh out to him that charmeth, nor is obedient to his saying—["the deaf adder that stoppeth her ears, and refuseth to hear the voice of the charmer, charm he never so wisely"]. This slaying Adder and venomous hath wit to love and affection, and loveth his make [mate] as it were by love of wedlock, and liveth not well without company. Therefore if the one is slain, the other pursueth him that slew that other with so busy wreak and vengeance that passeth weening. And knoweth the slayer, and reseth on him, be he in never so great company of men and of people, and busieth to slay him, and passeth all difficulties and spaces of ways, and with wreak of [will wreak] the said death of his make. And is not let ne put off, but it be by swift flight, or by waters or rivers. But against his malice kind giveth remedy and medicine. For kind giveth him right dim sight; for his eyes are set in the sides of his head, and be not set in the forehead; and therefore he may not see his adversary forthright, but aside. Therefore he may not follow his enemy by sight, but he followeth more by hearing and smell; for in these two wits he is strong and mighty. This Adder Aspis grieveth not men of Africa and Moors; for they take their children that they have suspect, and put them to these Adders: And if the children be of their kind this Adder Aspis grieveth them not: And if they be of other kind anon he dieth by venom of the Adder. These beasts slay strangers and men of other lands. And these serpents spare wonderly men that be born in the same land. So the serpent Anguis about the River Euphrates grieveth not nor hurteth men of the land; nor noyeth them that sleep, if they be of that land, and pain and slay busily other men, that be of other nations, what nation soever it be. Also Aristotle saith that in a certain mountain scorpions grieve no strangers; but they sting and slay men of the country.

Bartholomew (*Berthelet*), bk. xviii. § 9.

Asp's sting is not curable, but only with the water of a stone washed, which they take out of the sepulchre of an ancient king. *Batman's* addition to *Bartholomew*, *loc. cit.*

In Egypt so great is the reverence they bear to Asps, that if any in the house have need to rise in the night-time out of their beds, they first of all give out a sign by knacking of the fingers, lest they should harm the Asp, and so provoke it against them; at the hearing whereof, all the Asps get them to their holes and lodgings, till the person stirring be laid again in his bed. A domestical Asp had young ones; in her absence one of her young ones killed a child in the house; when the old one came again according to her custom to seek her meat, the killed child was laid forth, and so she understood the harm; then went she and killed that young one, and never more appeared in that house. Also there was an Asp that fell in love with a little boy that kept geese, whose love to the said boy was so fervent, that the male of the said Asp grew jealous thereof. Whereupon one day as he lay asleep, [he] set upon him to kill him, but the other seeing the danger of her love, awaked and delivered him. All the Asps of Nilus do thirty days before the flood remove themselves and their young ones into the mountains, and this is done yearly, once at the least. A man carrying a bottle of vinegar was bitten by an Asp, whiles by chance he trod thereupon, but as long as he bore the vinegar and did not set it down, he felt no pain thereby, but as often as to ease himself he set the bottle out of his hand, he felt torment by the poison.

Topsell, "History of Serpents," pp. 633-6.

Ass.

The Ass is a simple beast and a slow, and therefore soon overcome and subject to man's service. The elder the Ass is, the fouler he waxeth from day to day, and hairy and rough, and is a melancholic beast, that is cold and dry, and is therefore kindly heavy and slow, and unlusty, dull and witless and forgetful. Natheless he beareth burdens, and may away with travail and thraldom, and useth vile meat and little, and gathereth his meat among briars and thorns and thistles. Small birds that nesteth them in bushes, thorns, and briars hate the Ass. And therefore small sparrows fighteth with the Ass, for the Ass eateth the thorns, in the which the sparrows make their nests. And also the Ass rubbeth and froteth his flesh against the thorns, and so

the birds or the eggs of the sparrows falleth out of the nest down to the ground. And when that the Ass reareth and heaveth up her head, then by a strong blast the thorns moveth and shaketh, and of the great noise the birds be afeared full sore, and falleth out of the nest. And therefore the mothers suffereth them to leap on the face of the Ass, and bite and smite and rese to his eyes with their bills. And if the Ass have a wound or a scab in the ridge or in the side of pricking of thorns or in any other wise, the sparrows leapeth on the Ass, and pecketh with their bills in the wounds or in the sores, for the Ass should pass from their nests. And though such a sparrow be full little, yet unneath may the Ass defend himself against his rese, pricking and biting. The raven hateth full much the Ass, therefore the raven flyeth above the Ass, and laboureth with his bill to peck out his eyes; but the deepness of eyes helpeth the Ass, and thickness and hardness of the skin, for therewith the Ass closeth her eyes and heleth her sight, and defendeth against the resing and pricking of fowls. Also his long ears and moving thereof helpeth, for therewith he feareth small birds, that rese to peck out his eyes. The smoke of the Ass's hoof helpeth the birth of a child, in so much that it bringeth out a dead child, and shall not otherwise be laid to, for it slayeth a quick child if it be oft laid to, and lieth too long time. And new dirt of the same beast stauncheth blood wonderly. The Ass's milk, and Ass's blood helpeth against the biting of a scorpion. And men say, that if a man looketh in an Ass's ear when he is smit with a scorpion, anon the malice passeth. Also all venomous things fleeth smoke of the Ass's liver. Also the Ass's milk helpeth against venomous plaster, and against the malice of ceruse or of quicksilver. Also Ass's bones bruised and stamped and sod helpeth against venom, if the broth thereof be drunken. And urine of the male Ass with Nardus keepeth and saveth and maketh much hair. And the Ass dreadeth full sore to pass over water, and scrapeth therein; and the Ass passeth not gladly, where he may see the water through the planks, for he hath a feeble brain, and is soon grudged, and dreadeth therefore, and falleth through the chines of the bridge into the water, that he seeth running thereunder. And the Ass drinketh not gladly but of small wells that he is used to, and those that he may come dry-footed to. And wonder it is to tell, that though an Ass be sore athirst, if his water be changed, unneath he drinketh thereof, but if it be like the water that he is wont to drink of.

And the Ass hath another wretched condition known nigh to all men. For he is put to travail over night [might — *Bartholomew*], and is beaten with staves, and sticked and pricked with pricks, and his mouth is wrung with a barnacle [bit], and is led hither and thither, and withdrawn from leys and pasture that is in his way oft by refraining of the barnacle, and dieth at last after vain travails, and hath no reward after his death for the service and travail that he had living, not so much that his own skin is left with him, but it is taken away, and the carrion is thrown out without sepulchre or burials—but it be so much of the carrion that by eating and devouring is sometime buried in the wombs of hounds and wolves.

Bartholomew (*Berthelet*), bk. xviii. § 8.

When an Ass dieth, out of his body are engendered certain flies called Scarabees. Asses are subject to madness when they have tasted to certain herbs growing near Potnias. Some have used to put into gardens the skull of a mare or she-ass that hath been covered, with persuasion that the gardens will be the more fruitful. The wolf with small force doth compass the destruction of an Ass, for the blockish Ass, when he seeth a wolf, layeth his head on his side, that so he might not see, thinking that, because he seeth not the wolf, the wolf cannot see him.

Topsell, "Four-footed Beasts," pp. 19-21.

If a stone be bound to the tail of an Ass, he will not bray nor roar.

The skin of an Ass when it is hung over boys prevents them from being frightened.

If you wish that a man's head should appear as an Ass's head, take of the parings of [the hoof of] an Ass, and rub the man's head with them.

Albertus Magnus, "Of the Wonders of the World."

In Africa also are wild Asses, among which one male hath many females; a jealous beast, who (for fear of after encroaching) bites off the stones of the young males, if the suspicious female prevent him not by bringing forth in a close place, where he shall not find it.

Purchas' "Pilgrims," p. 558 (ed. 1616).

Baboon. *V.* Ape, Monkey.

You and your coach-fellow Nym . . . had looked through the grate, like a geminy of baboons.

Merry Wives of Windsor, ii. 2, 7-9.

Cool it with a baboon's blood,
Then the charm is firm and good.

Macbeth, iv. 1, 37-8.

BABOONS are a kind of apes, whose heads are like dogs, and their other parts like a man's. Some are much given to fishing; again, there are some which abhor fishes. Some there are which are able to write, and naturally to discern letters. They will eat venison, which they by reason of their swiftness take easily, and having taken it tear it to pieces, and roast it in the sun.

Topsell, "Four-footed Beasts," pp. 8, 9.

Balm.

Pierced to the soul with slander's venom'd spear,
The which no balm can cure but his heart-blood.

KING RICHARD II., i. 1, 171-2.

My pity hath been balm to heal their wounds.

3 KING HENRY VI., iv. 8, 41.

BALM drunk in wine is good against the bitings of venomous beasts, comforts the heart, and driveth away all melancholy and sadness. The juice thereof glueth together green wounds, being put into oil, unguent or Balm for that purpose, and maketh it of greater efficacy.

Gerard's "Herbal," *s.v.*

THIS Balm groweth in no place, but only there [*i.e.*, beside Cairo]. And though that men bring of the plants for to plant in other countries, they grow well and fair, but they bring forth no fructuous thing. And men cut the branches with a sharp flintstone or with a sharp bone, when men will go to cut them: for whoso cut them with iron, it would destroy his virtue and his nature. And men make always that Balm to be tilled of the Christian men or else it would not fructify, as the Saracens say themselves: for it hath been often time proved.

Sir John Mandeville, ch. v.

[He gives elaborate directions for distinguishing the true from the counterfeit balm.]

Balsam, Balsamum [*i.e.* Balm].

TIMON OF ATHENS, iii. 5, 110.
COMEDY OF ERRORS, iv. 1, 88.

BALSAMUM is set tofore all other smells, and was sometime granted but to one land among all lands, that is to wit Judea. And was not had nor found but in two gardens of the King's. *Bartholomew* (*Berthelet*), bk. xvii. § 18.

Barnacles.

We shall lose our time,
And all be turn'd to barnacles.
TEMPEST, iv. 1, 248-50.

I TOLD them of as great a marvel to them that is amongst us: and that was of the Barnacles. For I told them that in our country were trees that bear a fruit that become birds flying: and those that fall in the water live: and they that fall on the earth die anon: and they be right good to man's meat. And hereof had they a great marvel, that some of them trowed it were an impossible thing to be. *Sir John Mandeville*, ch. xxvi.

IN the Islands of Ireland, and Orcades, in certain places there, there be certain trees, much like unto willow-trees, out of which come forth certain little hairs, increasing by little and little into birds, having shape of ducks, hanging upon the bough by their nebs or bills; and when they are come to full perfectness, they fly away of themselves, or fall into the next seas, which birds we call Barnacles. This is related by the people that dwell there.

Lupton's "Notable Things," bk. vii. § 3.

[Gerard in his "Herbal" gives a description of the Barnacle or Goose-tree, too long to quote, but he declares that he has seen it, and vouches for it of his own knowledge.]

IN Man they have great store of Barnacles breeding upon their coasts. [He adds that he sought vainly for Barnacles until May, 1584, when he found many shells on ships in the Thames newly come home from Barbary or the Canary Isles, and on opening them he] saw the proportion of a fowl in one of them, saving that the head was not yet formed, because the fresh water had killed them all (as I take it). Certainly the feathers of the tail hang out of the shell at least two inches, the wings almost perfect, touching form, so that it cannot be denied but that some bird or other must proceed of this substance.

Harrison's "Description of Britain," p. 38, in *Holinshed*.

ONE little fish [Remora or Barnacle], not above half a foot long, is able to arrest and stay perforce, yea and hold as prisoners our goodly tall and proud ships. This little fish detained Caligula's ship (a galliass it was, furnished with five banks of oars to a side); so soon as ever the vessel was perceived alone in the fleet to stand still, presently they found one of these fishes sticking fast to the very helm. But this prince was most astonished at this, namely, That the fish sticking only to the ship should hold it fast, and the same, being brought into the ship, and there laid, not work the like effect. Neither do I doubt but all the sort of fishes are able to do as much.

Holland's Pliny, bk. xxxii. ch. 1.

Basilisk or Cockatrice.

Make me not sighted like the basilisk;
I have look'd on thousands, who have sped the better
By my regard, but kill'd none so.

WINTER'S TALE, i. 2. 388-90.

Come, basilisk,
And kill the innocent gazer with thy sight.

ii. KING HENRY VI., iii. 2, 52-3.

THE Cockatrice is a king of serpents, and they be afeard and flee when they see him. For he slayeth them with his smell and with his teeth; and slayeth also all thing that hath life, with breath and with sight. In his sight no fowl nor bird passeth harmless, and though he be far from the fowl, yet it is burnt and devoured by his mouth. But he is overcome of the weasel; and men bring the weasel to the Cockatrice' den where he lurketh and is hid. For the Father and Maker of all thing left no thing without remedy. And so the Cockatrice fleeth when he seeth the weasel, and the weasel pursueth and slayeth him. For the biting of the weasel is death to the Cockatrice; and never-

theless the biting of the Cockatrice is death to the weasel. And that is sooth, but if [unless] the weasel eat rue before. And against such venom, first the weasel eateth the herb of rue, though it be bitter, and by virtue of the juice of that herb, he goeth boldly and overcometh his enemy. And the Cockatrice is half a foot long, and hath white specks: And the Cockatrice slayeth that that he cometh nigh. As the scorpion he pursueth thirsty animals, and when they come to the water, he maketh them dropsical, and hydrophobic. For that water that he toucheth maketh the dropsy, and it is venomous and deadly. With hissing he slayeth, or he biteth or stingeth. And he presseth not his body with much bowing, but his course of way is forthright, and goeth in mean [the middle]. He dryeth and burneth leaves and herbs, not only with touch, but also by hissing and blast he rotteth and corrupteth all thing about him. And he is of so great venom and perilous, that he slayeth and wasteth him that nigheth him by the length of a spear, without tarrying; and yet the weasel taketh and overcometh him. And though the Cockatrice be venomous without remedy while he is alive, yet he loseth all the malice when he is burnt to ashes. *Bartholomew* (*Berthelet*), bk. xviii. § 8.

ITS head is very pointed, its eyes red, its colour inclining to black and yellow; it has a tail like a viper, but the rest of its body is like a cock. The Basilisk is sometimes gendered from a cock; for towards the end of summer a cock lays an egg from which the Basilisk is hatched. But many things must concur to this gendering, for it lays the egg in much warm dung, and there sits on it. And those who have seen its creation say that there is no shell to the egg, but a very strong skin which can resist the hardest blows. Also the opinion of some is that a viper or toad sits on that cock's egg—but this is doubtful.

Hortus Sanitatis, part iii. ("Of Birds") ch. xiii.

BASLE was built in the year 382, having the name of a Basilisk slain by a knight covered with crystal.

Fynes Moryson's "Itinerary," part i. ch. ii. p. 27.

EVEN as a lion is afraid of a cock, so is the Basilisk, for he is not only afraid at his sight, but almost dead when he heareth him crow. It is a question whether the Cockatrice die by the sight of himself. Once our nation was full of Cockatrices, and a certain man did destroy them by going up and down in glass, whereby their own shapes were reflected upon their own faces, and so they died. But this fable is not worth refuting, for it is more likely that the man should first have died by the corruption of the air from the Cockatrices.

Topsell, "History of Serpents," pp. 679, 681.

Bat.

All the charms
Of Sycorax, toads, beetles, bats, light on you.
TEMPEST, i. 2, 339-40.

Wool of bat, and tongue of dog.
MACBETH, iv. 1, 15.

THE reremouse [*i.e.* Bat] hating light flyeth in the even-tide with breaking and blenching and swift moving, with full small skin of her wings. And is a beast like to a mouse in sounding with voice, in piping and crying. And he is like to a bird, and also to a four-footed beast; and that is but seld found among birds. Reremice be blind as moles, and lick powder [dust] and suck oil out of lamps, and be most cold of kind; therefore the blood of a reremouse [a]nointed upon the eye-lids suffereth not the hair to grow again. *Bartholomew* (*Berthelet*), bk. xii. § 38.

IF you wish to see anything submerged and deep in the night, and that it may not be more hidden from thee than in the day, and that you may read books in a dark night,—anoint your face with the blood of a Bat, and that will happen which I say.

Albertus Magnus, "Of the Wonders of the World."

Bay, -tree.

Rosemary and bays.
PERICLES, iv. 6, 160.

The bay-trees in our country are all wither'd.
KING RICHARD, ii. 4, 8.

["Bay" was used in Shakespeare's time as a synonym for "laurel." *Cf. Minsheu's* Dictionary, *s.v.*, and *Cooper's* Thesaurus, *s.v. Laurus.*]

THIS tree worshippeth the house, and maketh it fair. The land that beareth laurel-tree is safe from lightning both in field and in house. *Bartholomew* (*Berthelet*), bk. xvii. § 48.

BAY-BERRIES taken in wine are good against the bitings of any venomous beast, and against all venom and poison. The oil pressed out of these cureth them that are beaten black and blue, and that be bruised by squats and falls. Common drunkards were accustomed to eat in the morning fasting two leaves thereof against drunkenness.

Gerard's "Herbal," bk. iii. ch. lxviii.

In the year 1629 at Padua, preceding a great pestilence, almost all the Bay-trees about that famous university grew sick and died. *Evelyn's* "Sylva," bk. ii. ch. vi.

Beagle. *V.* Brach.

TWELFTH NIGHT, ii. 3, 195.
TIMON OF ATHENS, iv. 3, 175.

Bean.

Peas and beans are as dank here as a dog, and that is the next way to give poor jades the bots.

1 KING HENRY IV., ii. 1, 9-10.

A fat and bean-fed horse.
MIDSUMMER NIGHT'S DREAM, ii. 1, 45.

The Bean is a manner codware, and serveth to pottage, and in old time men used to eat thereof. Beans cause vain dreams and dreadful. Many meddle beans with bread-corn, to make the bread the more heavy. By oft use thereof the wits be dulled. Or else, dead men's souls be therein. Therefore the bishop should not eat Beans. Beans grow in Egypt with sharp pricks, therefore crocodiles flee from them, and dread lest their eyes should be hurt with the sharp pricks of them. Such a Bean is x cubits long, with a head as a poppy, and therein Beans be closed, and that head is red as a rose. And those Beans grow not on stalks nor in cods. *Bartholomew* (*Berthelet*), bk. xvii. § 64.

The skins of Beans applied to the place where the hairs were first plucked up, will not suffer them to grow big, but rather consumeth their nourishment.

Gerard's "Herbal," bk. ii. ch. v. and vii.

In June buttered Beans saveth fish to be spent.
Tusser, "A Hundreth Good Poyntes of Husbandrie."

You may imagine it to be Twelfth-day at night, and the Bean found in the corner of your cake.

Rowley's "Woman Never Vexed," ii. 1.

Now, now the mirth comes
With the cake full of plums.
Where Bean's the king of the sport here.

Herrick's "Hesperides."

THE choosing of a person King or Queen by a bean found in a piece of a divided cake was formerly a common Christmas gambol in both the English universities.

Brand's "Popular Antiquities," vol. i. p. 20.

[See also the same author, p. 97, under "Mid-Lent Sunday." French Beans are mentioned in *Beaumont and Fletcher's* "Tragedy of Bonduca" (i. 2).]

SHE made me colour my hair with Bean-flower to seem elder than I was. *Webster's* "Devil's Law Case," iv. 2.

Bear.

Thy groans
Did make wolves howl and penetrate the breasts
Of ever-angry bears.

TEMPEST, i. 2, 287-9.

Wolves and bears, they say,
Casting their savageness aside, have done
Like offices of pity.

WINTER'S TALE, ii. 3, 187-9.

The rugged Russian bear.

MACBETH, iii. 4, 100.

Like to a chaos or an unlick'd bear-whelp
That carries no impression like the dam.

3 KING HENRY VI., iii. 2, 161-2.

One bear will not bite another.

TROILUS AND CRESSIDA, v. 7, 19.

Unicorns may be betray'd with trees,
And bears with glasses.

JULIUS CÆSAR, ii. 1, 204-5.

AMID the desert rocks the mountain bear
Brings forth unform, unlike herself, her young;
Nought else but lumps of flesh withouten hair.
In tract of time her often licking tongue
Gives them such shape, as doth (erelong) delight
The lookers on.

Arthur Broke's "Romeus and Juliet," *Address to the Reader.*

WHEN the Bear cannot find origanum to heal his grief, he blasteth all other leaves with his breath.

Lilly's "Sappho and Phaon" (Prologue).

Their gendering is in the beginning of winter, and gender not as other fourfooted beasts do, but they gender both lying, and then they depart asunder each from other, and go in dens either by themself, and whelpeth therein the xxx day, and the whelps be not more than five, and be white and evil shapen. For the whelp is a piece of flesh little more than a mouse, having neither eyes nor hair, and having claws somedeal bourging [*i.e.*, burgeoning], and so this lump she licketh, and shapeth a whelp with licking. And so men shall see no where beasts more selder gender nor whelp than Bears, and therefore the males hide them and lurk forty days, and the females array their houses four months with boughs, fruit and branches, and covereth it, for to keep out the rain with nesh twigs and branches. The first forty days of these days they sleep so fast, that they may not be awaked with wounds, and that time they fast mightily. And the grease of a Bear helpeth against the falling of the hair. And after these days, she sitteth up and liveth by sucking of her feet, and beclippeth the cold whelps, and holdeth them fast to her breast: And heateth and comforteth them, and lieth grovelling upon them, as birds do. And it is wonder to tell a thing that Theophrastus saith and telleth, that Bear's flesh sod that time vanisheth if it be laid up, and is no token of meat found in the almery [cupboard, larder], but a little quantity of humour: and hath that time small drops of blood about the heart, and no manner of blood in the other deal of the body. And in springing time the males go forth and be fat, and the cause thereof is unknown, namely for that time they be not fatted with meat neither with sleep, but only seven days. And when she goeth out of her den, she seeketh an herb, and eateth it to make lax her womb, that is then hard and bound. Then her eyes be dimmed, and therefore namely they labour to get them honey-combs, for the mouth should be wounded with stinging of bees and bleed; and so relieve the heaviness and sore ache of their eyes. His head is full feeble, that is most strong in the lion, and therefore sometime he falleth down headlong upon the rocks, and falleth upon gravel and dieth soon. And as men say, the Bear's brain is venomous, and therefore when they be slain, their heads be burnt in open places, for men should not taste of the brain, and fall into woodness of Bears. And no beast hath so great sleight to do evil deeds as the Bear. And the Bear eateth crabs and ants for medicine, and eateth flesh for great strength, and is an unpatient beast and wrathful, and will be avenged on all those that him toucheth. If another touch him, anon he leaveth the first, and reseth on the second, and reseth on the third; and when he is taken, he is made blind with a bright basin [*cf.* quotation from "Julius Cæsar"] and is bound with chains, and compelled to play: and tamed with beating, and is an unsteadfast beast and unstable, and uneasy, and goeth therefore all day about the stake to which he is strongly tied. He licketh and sucketh his own feet, and hath liking in the juice thereof. He can wonderly stie [climb] upon trees unto the highest tops of them [and robs wild bees of their honey]. And the hunter taketh heed thereof, and pitcheth full sharp hooks and stakes about the foot of the tree, and hangeth craftily a right heavy hammer or a wedge tofore the open way to the honey, and then the Bear cometh, and is an hungered, and the log that hangeth there on high letteth him, and he putteth away the wedge dispiteously, but after the removing, the wedge falleth again and hitteth him on the ear, and he hath indignation thereof; and putteth away the wedge dispiteously and right fiercely, and then the wedge falleth and smiteth him harder than it did before, and he striveth so long with the wedge, until his feeble head doth fail by oft smiting of the wedge, and then he falleth down upon the pricks and stakes, and slayeth himself in that wise. Bears licketh not drink, as beasts do with sawy teeth; and sucketh not neither swalloweth, as beasts do that have continual teeth, as sheep and men; but biteth the water and swalloweth it. *Bartholomew (Berthelet)*, bk. xviii. §§ 112-3.

Beast. *V.* Animal.

Bee.

Like the bee, culling from every flower
The virtuous sweets,
Our thighs pack'd with wax, our mouths with honey,
We bring it to the hive, and, like the bees,
Are murdered for our pains.
ii. King Henry IV., iv. 5, 75-81.

'Tis seldom when the bee doth leave her comb
In the dead carrion.

ii. KING HENRY IV., iv. 4, 79-80.

So work the honey bees,
Creatures that by a rule in nature teach
The act of order to a peopled kingdom.
They have a king and officers of sorts;
Where some, like magistrates, correct at home, *etc.*

KING HENRY V., i. 2, 187-204.

The commons, like an angry hive of bees,
That want their leader, scatter up and down,
And care not who they sting in his revenge.

ii. KING HENRY VI., iii. 2, 125-8.

BEES be cunning and busy in office of making of honey, and they dwell in their own places that are assigned to them, and challenge no place but their own. And they build and make their houses with a wonderful craft, and of divers flowers; and they make honey-combs, wound and writhen with wax full craftily, and fill their castles with full many children. They have an host and a king, and move war and battle, and fly and void smoke and wind, and make them hardy and sharp to battle with great noise. Many have assayed and found that often Bees are gendered and come of carrions of rothern [*i.e.*, cattle]. And for to bring forth Bees, flesh of calves which be slain is beat that worms may be gendered and come of the rotted blood, the which worms after take wings and be made Bees, as shern-birds [*i.e.*, hornets] be gendered of carrions of horses. Bees make among them a king, and ordain among them common people. And though they be put and set under a king, yet they be free and love their king that they make by kind love; and defend him with full great defence; and hold honour and worship to perish and be spilt for their king; and do their king so great worship that none of them dare go out of their house, nor to get meat, but if the king pass out and take the principality of flight. And Bees choose to their king him that is most worthy and noble in highness and fairness, and most clear in mildness, for that is chief virtue in a king. For though their king have a sting, yet he useth it not in wreak. And kindly the more huge Bees are, the more lighter they be, for the greater Bees be lighter than the less Bees. And also Bees that are unobedient to the king, they deem themselves by their own doom for to die by the wound of their own sting. Also Bees sit upon the hives and suck the superfluity that is in honey-combs. And it is said that if they did not so, thereof should attercops [*i.e.*, spiders] be gendered of that superfluity, and the Bees should die.

Bartholomew (*Berthelet*), bk. xii. § 4.

If the night falleth upon them in their journey, then they lie upright to defend their wings from rain and from dew, that they may in the morrow tide fly the more swifter to their work with their wings dry and able to fly. And they ordain watches after the manner of castles, and rest all night until it be day, till one Bee wake them all with twice buzzing or thrice, or with some manner trumping: then they fly all, if the day be fair on the morrow. And the Bees that bringeth and beareth what is needful, dread blasts of wind, and fly therefore low by the ground when they be charged, lest they be letted with some manner of blasts; and chargeth themself sometime with gravel or with small stones, that they may be the more steadfast against blasts of wind by heaviness of the stones. Bees be comforted with smell of crabs, if they be sodden nigh them. They die all with oil as such round beasts do, and namely if the head be anointed; and such beasts, set in the sun, quicken again if they be bespring with vinegar. And Bees that make honey slay the males that grieve them, and evil kings, that rule them not aright, but only eat too much honey. And no creature is more wreakful, nor more fervent to take wreak than is the Bee when he is wroth; therefore a multitude of the host of Bees throweth down great hedges when they be compelled to withstand them that destroy their honey. And Bees be pleased with harmony and melody of sound of song, and with flapping of hands and beating of basins. And therefore with beating of basins, tinging and tinking of timbers, they be comforted and called to the hives. *Ibid.*, bk. xviii. § 12.

Where the Bee can suck no honey, she leaveth her sting behind. *Lilly*, "Sappho and Phaon" (Prologue).

Flies that die on the honeysuckle become poison to Bees. *Ibid.*, ii. 4.

A Bee's sting pricketh deepest, when it is fullest of honey. *Ibid.*, iv. 4.

Beetle.

The poor beetle that we tread upon,
In corporal sufferance, finds a pang as great
As when a giant dies.

MEASURE FOR MEASURE, iii. 1, 79-81.

The shard-borne beetle, with his drowsy hums,
Hath rung night's yawning peal.

MACBETH, iii. 2, 42-3.

BEETLES are often produced from the putrid flesh of horses. They are hung round the necks of infants for their cure. The nature of the green Beetle sharpens the sight of those who behold it, and therefore carvers of jewels take pleasure in the sight of it.

Hortus Sanitatis, part iii. ("Of Birds"), ch. cvi. (translated).

THE Beetle is bred of putrid things and of dung, and it chiefly feeds and delights in that. Of all plants they cannot away with rose-trees, for they die by the smell of them. They have no females, but have their generation from the sun. Though the eagle, its proud and cruel enemy, do make havoc and devour this creature of so mean a rank, yet as soon as it gets an opportunity it returneth like for like. For it flieth up nimbly into her nest with its fellow-soldiers the scarab-beetles, and in the absence of the old she-eagle bringeth out of the nest the eagle's eggs one after another, which, falling and being broken, the young ones are deprived of life.

Mouffet, "Theatre of Insects," pp. 1005-13.

Bell-wether. *V.* Wether.

A jealous rotten bell-wether.

MERRY WIVES OF WINDSOR, iii. 5, 111.

Benedictus (Carduus).

MARG. Get you some of this distilled Carduus Benedictus, and lay it to your heart; it is the only thing for a qualm.

HERO. There thou prickest her with a thistle.

BEAT. Benedictus! why Benedictus? You have some moral in this Benedictus.

MARG. Moral! no, by my troth, I have no moral meaning; I meant plain Holy thistle.

MUCH ADO ABOUT NOTHING, iii. 4, 73-80.

CARDUUS is a manner herb or a weed with pricks. The kind thereof is biting and cruel. Therefore the juice thereof cureth the falling of the hair. The root thereof sod in water giveth appetite to drinkers, and is most profitable to the mother, and therefore it is no wonder though women desire it. And in drawing up of carduus men's fingers be oft grieved with pricks.

Bartholomew (*Berthelet*), bk. xvii. § 36.

CARDUUS BENEDICTUS is diligently cherished in gardens in these Northern parts. [It is called] in English Blessed Thistle, but more commonly by the Latin name Carduus Benedictus. Blessed Thistle taken in meat or drink is good for the swimming and giddiness of the head, it strengtheneth memory and is a singular remedy against deafness. The juice of the said Carduus is singular good against all poison.

Gerard's "Herbal," *s.v.* See also *Lupton's* "Notable Things," bk. ii. § 84, and bk. iv. § 53.

Bilberry.

Where fires thou find'st unraked, and hearths unswept,
There pinch the maids as blue as bilberry.

MERRY WIVES OF WINDSOR, v. 5, 48-9.

[Bilberries (*Vaccinium myrtillus*) are identified by Gerard with worts (*V. uliginosum*) or whortleberries, and he says that the red worts have purple berries, and that the people of Cheshire do eat the black whortles in cream and milk (bk. iii. ch. lxxiii.) —as is done in the West of England at this day.]

Birch.

As fond fathers,
Having bound up the threatening twigs of birch,
Only to stick it in their children's sight
For terror, not to use, in time the rod
Becomes more mock'd than fear'd.

MEASURE FOR MEASURE, i. 3, 23-7.

IN times past the Magistrates' rods were made hereof; and in our time also the schoolmasters and parents do terrify their children with rods made of Birch. It serveth well to the decking up of houses, and banqueting rooms, for places of pleasure, and beautifying of streets in the cross or gang week, and such like.

Gerard's "Herbal," bk. iii. ch. cxiv.

BIRCH hath many hard twigs and branches with knots, and therewith often children be chastised and beaten on the bare buttocks and loins. And of the boughs and branches thereof be besoms made to sweep and to cleanse houses of dust and of other uncleanness. And this tree hath much sour juice and somewhat biting. And men use therefore in springing time and in harvest to drink it in stead of wine but it feedeth not, nor nourisheth not, nor maketh men drunk. *Bartholomew* (*Berthelet*), bk. xvii. § 159.

Bird.

THE crane that walketh for the watch by night, holdeth a little stone in his foot, that if he hap to fall asleep, he may be waked by falling of the stone.

Bartholomew, *Berthelet*, bk. xii., Introduction.

To take the Birds that eat the seeds that are sown; seethe garlick that it may not grow again; for it is said to profit marvellously, if it be thrown unto them; for they that shall eat of it, will be taken with your hand.

If you will make Birds drunk that you may catch them with your hands, take such meat as they love, as wheat or beans, or such like, and lay the same to steep in lees of wine, or in the juice of hemlocks, and sprinkle the same in the place where the Birds use to haunt; and if they do eat thereof, straightways they will be so giddy, that you may take them with your hands. I wrote this out of an old written book, wherein I know many true things were written. *Lupton's* "Notable Things," bk. viii. §§ 4 and 68.

If you wish to understand the speech of Birds, take with you two friends on the fifth day of the Calends of November, and go into a grove with your dogs as if to hunt, and take the first beast you find home with you, and prepare it with the heart of a fox, and straightway you will understand the speech of Birds or beasts; and if you desire that any one else should understand it,—kiss him, and he will understand likewise.

Albertus Magnus, "Of the Wonders of the World."

Of such wild fowl as are bred in our land, we have the crane, the bittern, the wild and tame swan, the bustard, the heron, curlew, snite [snipe], wild-goose, wind or dotterel, brant [brant-goose or barnacle], lark, plover of both sorts, lapwing, teal, widgeon, mallard, sheldrake, shoveler, peewit, seamew, barnacle, quail (who only with man are subject to the falling sickness), the knot, the oliet or olife, the dunbird, woodcock, partridge and pheasant, besides divers other. As for egrets, pawpers and such like, they are daily brought to us from beyond the sea. Our tame fowl are common both to us and to other countries, as cocks, hens, geese, ducks, peacocks of Ind, pigeons. I would likewise entreat of other fowls which we repute unclean, as ravens, crows, pies, choughs, rooks, kites, jays, ring-tails, starlings, wood-spikes, woodgnaws, etc. Our other fowls are nightingales, thrushes, blackbirds, mavises, ruddocks, redstarts or dunnocks, larks, tivits, kingfishers, buntings, (turtles, white or grey), linnets, bulfinches, goldfinches, wash-tails, cherry-crackers, yellowhammers, fieldfares, etc.

Harrison's "Description of England," pp. 222-3, in *Holinshed.*

Bird-lime.

My invention
Comes from my pate as bird-lime does from frize.
OTHELLO, ii. 1, 126-7.

The glue which is made of the berries of mistletoe is called Bird-lime. *Gerard's* "Herbal."

THRUSHES eat the berries, and roost all night on the mistletoe-tree, and by their sitting and [cacando] the mistletoe beareth Bird-lime, the bane of the bird.

Minsheu's Dictionary, *s.v.* "*Mistletoe.*"

Bitch. *V.* Dog.

MERRY WIVES OF WINDSOR, iii. 5, 11.

Bitter-sweeting.

Thy wit is a very bitter-sweeting; it is a most sharp sauce.
ROMEO AND JULIET, ii. 4.

[The commentators will have "Bitter-sweeting" to be an apple, and quote in proof instances of the word "Bitter-sweet," which Gerard in his "Herbal" identifies with the woody nightshade. "Bitter-sweet" or "Bitter-scale" is mentioned as a Dorsetshire apple in John Newburgh's "Observations concerning Cider," quoted in Evelyn's "Pomona."]

Blackberry.

If reasons were as plentiful as blackberries, I would give no man a reason upon compulsion.
i. KING HENRY IV., ii. 4, 264-6.

Shall the blessed sun of heaven prove a micher and eat blackberries?
i. KING HENRY IV., ii. 4, 449-50; also
TROILUS AND CRESSIDA, v. 4, 12.

[Gerard in his "Herbal" classes the raspberry and the knot-berry (or cloud-berry) with the bramble or Blackberry. He says:]

THE bramble groweth for the most part in every hedge and bush. Bk. iii. ch. 4.

ON Michaelmas-day the devil puts his foot upon the Blackberries. *Notes and Queries.*

Blind-worm.

Newts and blind-worms, do no wrong,
Come not near our fairy queen.
MIDSUMMER NIGHT'S DREAM, ii. 2, 11-12

Adder's fork and blind-worm's sting.
MACBETH, iv. 1, 16.

[Also called] sloe-worm, because it useth to creep and live on sloe-trees. *Minsheu's* Dictionary, *s.v.*

It is small, and has no eyes. *Hortus Sanitatis*, ch. xxxvi.

THE Blindworm is sometimes confounded with the *amphisbœna*, a serpent with two heads, one in the usual place, the other at the end of its tail, and moving either way. This serpent is the first to appear, being anxious about its eggs. While one part of it keeps watch, the other sleeps; and its eyes shine like lanterns. There is another that walks upon its heels, and upon its tail.

Chiefly from the *Hortus Sanitatis*, ch. ix.

Bloodhound.

You starved bloodhound.
ii. KING HENRY IV., v. 4, 31.

HOUNDS pursue the foot of prey by smell of blood.
Bartholomew (*Berthelet*), bk. xviii. § 25.

THERE is a certain class of hounds which know thieves by the smell; and with implacable hatred distinguish them from other men. *Hortus Sanitatis*, ch. xxiv.

Blood-sucker (*i.e.*, a Leech—*Minsheu's* Dictionary, *s.v.*).

Pernicious blood-sucker of sleeping men.
ii. KING HENRY VI., iii. 2, 226.

A LEECH sitteth upon venomous things, and therefore when he shall be set to a member because of medicine, first he shall be wrapped in nettles and in salt, and is thereby compelled to cast out of his body if he hath tasted any venomous thing in warm water.

Bartholomew (*Berthelet*), bk. xviii. § 93.

IT has neither bones, feet nor wings. By sucking too much blood, it often causes its own death. It draws out putrid blood, and kills itself while healing its victim.

Hortus Sanitatis, ch. cxxxi.

Box-tree.

Get ye all three into the box-tree.

TWELFTH NIGHT, ii. 5, 18.

BOX holdeth long time shapes and figures which be made therein; so thereof be made fair images and long-during. The shaving of Box dyeth hair that is oft washen in the broth thereof.

Bartholomew (*Berthelet*), bk. xvii. § 20.

FOOLISH empirics and women leeches do minister it against the apoplexy and such diseases. Turners and cutlers, if I mistake not the matter, do call this wood dudgeon, wherewith they make dudgeon-hafted daggers.

Gerard's "Herbal," bk. iii. ch. lxx.

THE leaves and the dust of the wood boiled in lye will make hairs of an auburn (or Abraham) colour. I learned of a friend who had tried it effectual, to cure the biting of a mad dog—take the leaves and roots of cowslips, of the leaves of Box and penny-royal, of each a like quantity, shred them small, and put them into hot broth, and let it be so taken three days together, and apply the herb, etc., to the bitten place with soap and hogs' suet melted together.

Parkinson's "Herbal," *s.v.*

Box-combs bear no small part
In the militia of the female art;
They tie the links which hold our gallants fast
And spread the nets to which fond lovers haste.

The oil assuages the tooth-ache. But the honey which is made at Trebizond in Box-trees, renders them distracted who eat of it.

Evelyn's "Sylva," bk. ii. ch. vi.

Brake.

Through bog, through bush, through brake, through brier.
MIDSUMMER NIGHT'S DREAM, iii. 1, 110.

Brake is the female fern.
Gerard.

A BRAKE of fern, because wild beasts break out of them.
Minsheu's Dictionary

Brach.

Brach Merriman the poor cur is emboss'd;
And couple Clowder with the deep-mouth'd brach.
TAMING OF THE SHREW, Induction 1, 17-8.

["Brach" is defined in Minsheu's Dictionary as "a little hound," and the Italian equivalent given is *Bracca,* which Florio in his Dictionary gives "a brach, a bitch, a beagle." In the last sense "brach" is used in "King Lear" (iii. 6, 71-2):

Mastiff, greyhound, mongrel grim,
Hound or spaniel, brach or lym.

"Brach," *i.e.*, bitch, occurs in "i. King Henry IV.," iii. 1, 241:

I had rather hear Lady my brach howl in Irish;

and in "King Lear," i. 4, 125; *cf.* Nares' Glossary.]

Bramble. *V.* Blackberry.

AS YOU LIKE IT, iii. 2, 380.

Is dark and shadowy by reason of his thickness and is therefore friend to adders and other creeping worms. Therefore it is not sicher to sleep and rest nigh such bushes for such venomous worms.
Bartholomew (*Berthelet*), bk. xvii. § 40.

Brass—Brazen.

Pewter and brass and all things that belong
To housekeeping.
TAMING OF THE SHREW, ii. 1, 356-7.

BRASS and copper be called Æs, for either is made of the same stone by working of fire, for a stone resolved with heat turneth into Brass. Brass and copper be made in this manner as other metal be of brimstone and quicksilver, and that happeth when there is more of brimstone than of quicksilver. If Brass be meddled with other metal, it changeth both colour and virtue, as it fareth in latten. Brazen vessels be soon red and rusty, but they be oft scoured with sand, and have an evil savour and smell but they be tinned. Also Brass, if it be without tin, burneth soon.

Bartholomew (*Berthelet*), bk. xv. § 37.

RICHMONDSHIRE—the mountains plentifully yield lead, pit-coals and some Brass. . . . Cumberland hath mines of Brass [*i.e.*, copper].

Fynes Moryson, "Itinerary," part iii., p. 144.

Breese.

Yon ribaudred nag of Egypt,

* * * * *

The breese upon her, like a cow in June,
Hoists sails and flies.

ANTONY AND CLEOPATRA, iii. 10, 10-5.

In her ray and brightness
The herd hath more annoyance by the breese
Than by the tiger.

TROILUS AND CRESSIDA, i. 3, 47-9.

The horrid Breese man's body doth not spare,
He flies from us into the open air.

But they fled home as herds of oxen do,
When that the Breese doth force them for to go,
In the springtime when days do longer grow.

THE fly called *æstrum* is of a yellowish colour, who when it enters the ears of an ox causeth him to run mad; he carries before him a very hard, stiff and well-compacted sting, with which he strikes through the ox his hide. They follow oxen and horses and young cattle by scent of their sweat, because they cannot reach them with their sight, being very weak-sighted. They are generated of the worms that come out of the wood putrefied [or, according to another authority, from horse-leeches].

Mouffet, "Theatre of Insects," pp. 935-6.

Briar. *V.* Rose.

TIMON OF ATHENS, iv. 3, 422.

THE root of the Briar-bush is a singular remedy found out by oracle against the biting of a mad dog. The fruit when it is ripe maketh most pleasant meats and banqueting dishes, as tarts and such like; the making whereof I commit to the cunning cook, and teeth to eat them in the rich man's mouth. *Gerard's* "Herbal," bk. iii. ch. iii.

Brimstone. *V.* Sulphur.

To put fire in your heart, and brimstone in your liver.
TWELFTH NIGHT, iii. 2, 21-2.

IF you would have any beast or any part of the same (of what colour soever he be) to be turned into white, shave off the hairs, and smoke the same that is shaven with the fume of Brimstone, and white hairs will grow there. You may prove the same in flowers.

Lupton's "Notable Things," bk. vi. § 1.

Brock [Badger].

TWELFTH NIGHT, ii. 5, 114.

THE Brock is a beast of the quantity of a fox, and his skin is full hairy and rough. In such beasts is wit and flight, and holdeth in the breath, and blowing; [and] stretcheth the skin so holding their breathings, when they be hunted and chased with hunters' dogs, and so they find sleight and manner, by such strutting out of the skin, to eschew and put off the biting of those hounds that so do pursue and follow to noy them, and also for to slay them, and in like wise put they off the smitings of the hunters. These beasts know when tempest shall fall, and maketh them therefore dens under earth with diverse enterings, and when the Northern wind bloweth, he stoppeth the north entering with his rough tail, and letteth stand open the south entering, and againwards. There is a manner kind of

Brocks that gather meat with the female against winter, and layeth it up in his den, and when cold winter cometh, the male dreadeth lest store of meat should fail, and refraineth the female, and withdraweth her meat and suffereth her not to eat her fill, and she feigneth peace, as it were following the male's will, and cometh in on that other side of the den, and openeth her jaws, and eateth and devoureth and wasteth the meat that is gathered, unwitting the male. These beasts hate the fox, and fight oft-times with him, but when the fox seeth that he may not for roughness and for hardness of the skin grieve him, he feigneth him as though he were sick and overcome, and fleeth away, and while the Brock goeth out to get his prey, the fox cometh into his den, and defileth his chamber with urine and other uncleanness. And the Brock is squeamish of such foul things, and forsaketh his house that is so defiled, and getteth needfully another dwelling-place.

Bartholomew (*Berthelet*), bk. xviii. § 103.

THE Brock has short legs, and not equal on the two sides, but shorter on the left side, so that planting the feet of the right side in the ruts made by wheels, it runs valiantly, and escapes its pursuers. The fat of the Badger grows when the moon waxes, and decreases as it wanes, so that if it be killed on the last day of the old moon none is found. This is strange, that though this part of the beast is medicinal, yet its bite is often very serious and fatal; and the reason of this is that it lives on wasps, and animals which creep on the ground, and are venomous, and therefore they infect its teeth. Its brain boiled with oil cures all pains. *Hortus Sanitatis*, part ii. ch. cxlii.

[Sir Toby probably calls Malvolio "Brock" in allusion to the habit described by Bartholomew of this animal in strutting (puffing) out its skin, so the word conveys a vivid and ludicrous idea of Malvolio's gait.]

WE have Badgers in our sandy and light grounds, where woods, furzes, broom and plenty of shrubs are to shrowd them in, when they be from their burrows. Foxes and Badgers are rather preserved by gentlemen to hunt and have pastime withal at their own pleasures, than otherwise suffered to live, as not able to be destroyed because of their great numbers.

Harrison's "Description of England," p. 225 (1586), in *Holinshed.*

Buck. *V.* Hart, Stag, Deer.

Bugle.

[Bugle-Bracelet is probably a bracelet of glass beads ("Winter's Tale," iv. 4, 224), but "your Bugle eyeballs" ("As You Like It," iii. 5, 47) may refer to the Bugle or buffalo, as "Bugle-browed" in *Middleton's* "Anything for a Quiet Life." Phebe quotes Rosalind's words with a difference in l. 130:

He said mine eyes were black.

Bartholomew (bk. xviii. § 15) describes the Bugle (*i.e.*, buffalo) as black or red. Or "Bugle eyeballs" may have a similar meaning to Homer's "ox-eyed."]

BUGLE flesh sod or roasted healeth man's biting. His marrow taken out of the right leg doth away hair off the eyelids. His hoof with myrrh fasteneth wagging teeth. And Bugle-milk is full good against smiting of serpents and of scorpions, and against venom of the cricket [and of the salamander]. Also some be wonderful great, and nevertheless most quiver and swift; in so much *ut fimum quem projiciunt* in turning about falleth on their horns or ever it may come to the ground. When the cow's time of calving cometh, many of them come about her, and make of dirt as it were a wall. *Bartholomew, ut supra.*

Bull.

BULLS of Ind be red, and swift and cruel, and their hair is turned in contrary wise, and such a Bull bendeth the neck at his own will, and putteth off darts and shot with hardness of the back; and is fierce and is not overcome; and when he is tied under a fig-tree, he loseth and leaveth all his fierceness, and is suddenly sober and soft. If thou dost cut and slit his skin, so that it arear somewhat from his flesh with blowing with a pipe, and givest him afterward to eat, then he fatteth; and is made fat with sweetmeats, as with figs and grapes and raisins. Some Bulls have movable horns, and move them one after another in fighting; and be always fierce when they be taken, and destroy themselves, and die for indignation.

Bartholomew (*Berthelet*), bk. xviii. § 100.

IF the right knee of a Bull be tied with a broad band, it will make him tame.

Lupton, "A Thousand Notable Things," bk. iii. § 64.

A BULL is the husband of a cow, and ringleader of the herd. When Bulls fight with wolves, they wind their tails together, and so drive them away with their horns. The blood of Bulls is accounted among the chiefest poisons.

Topsell, "Four-footed Beasts," pp. 47-50.

[IN Cibola, near Mexico,] they drink the blood of the ox hot (which of our Bulls is counted poison).

Purchas' "Pilgrims," p. 778 (ed. 1616).

Bullock. *V.* Bull.

Bunting.

My dial goes not true; I took this lark for a bunting.
ALL'S WELL THAT ENDS WELL, ii. 5, 6-7.

The goss-hawk beats not at a bunting.
Ray's "Proverbs."

[The Bunting is the woodlark.]

Burnet.

The freckled cowslip, burnet, and green clover.
KING HENRY V., v. 2, 49.

BURNET is a singular good herb for wounds; it stauncheth bleeding, as well inwardly taken, as outwardly applied. The lesser Burnet is pleasant to be eaten in salads, in which it is thought to make the heart merry and glad, as also being put into wine, to which it yieldeth a certain grace in the drinking. *Gerard's* "Herbal," *s.v.*

[Evelyn, in his "Acetaria, or Discourse of Sallets," gives the same characteristics of Burnet.]

Burr.

I am a kind of burr; I shall stick.
MEASURE FOR MEASURE, iv. 3, 189-90.

They are but burrs, cousin, thrown upon thee in holiday foolery; if we walk not in the trodden paths, our very petticoats will catch them.
AS YOU LIKE IT, i. 3, 13-6.

[BURR] the Clete groweth by old walls; and hight Philanthropos, as it were loving mankind, for it cleaveth to men's clothes by a manner affection and love, as it seemeth. They heal smiting of scorpions, nor they smite not a man that is balmed with the juice thereof.

Bartholomew (*Berthelet*), bk. xvii. § 93.

THE Burr or fruit of the lesser Burr dock before it be fully withered, being stamped and put into an earthen vessel, and afterwards when need requireth the weight of two ounces thereof and somewhat more, being steeped in warm water and rubbed on, maketh the hairs of the head red; yet the head is first to be dressed or rubbed with nitre. The roots being stamped with a little salt, and applied to the biting of a mad dog, cureth the same, and so speedily setteth free the sick man. The juice of the leaves drunk with old wine doth wonderfully help against the bitings of serpents. The stalk of Clot burr before the Burrs come forth, the rind pilled off, being eaten raw with salt and pepper, or boiled in the broth of fat meat, is pleasant to be eaten. *Gerard's* "Herbal," *s.v.*

Butterfly.

Butterflies
Show not their mealy wings but to the summer.
TROILUS AND CRESSIDA, iii. 3, 78-9.

There is differency between a grub and a butterfly; yet your butterfly was a grub.
CORIOLANUS, v. 4, 11-2.

BUTTERFLIES are small birds, which chiefly abound when the mallows are in flower. Butterflies are flying grubs, which get their food from flowers. The female lays eggs, and dies after laying them; the eggs last through the winter, and in the summer become grubs, which, invigorated by the warmth of the sun and by nocturnal dew, produce wings for flying. Butterflies should be killed in the month of April when they hurt the bees.

Hortus Sanitatis, part iii. ("Of Birds") § 96.

BUTTERFLIES be called small fowls, and be most in fruit in apples, and breedeth therein worms that come of their stinking filth. For of malshrags [caterpillars] cometh and breedeth Butterflies, and of the dirt of Butterflies left upon leaves breedeth and cometh again malshrags.

Bartholomew (*Berthelet*), bk. xviii. § 47.

[In Mouffet's "Theatre of Insects" are described and pictured some eighty different moths and Butterflies (including apparently, some flies and beetles), but no English names are given. He says that the venomous dung of Butterflies, with aniseed, goat's milk cheese, hog's blood, galbanum, and opoponax made into troches (or lozenges) with good sharp wine, and dried in the sun, allure fish to your hook.]

Buzzard.

O slow-wing'd turtle! shall a buzzard take thee?
TAMING OF THE SHREW, ii. 1, 208.

More pity that the eagle should be mew'd
While kites and buzzards prey at liberty.
KING RICHARD III., i. 1, 132-3.

THE Buzzard is of the class of hawks; but somewhat darker, and very slow and sluggish in flight; yet it lives on prey, which it is able to catch by cunning, or when it is let by some sickness or slowness. This bird is very sweet in taste. *Hortus Sanitatis*, ch. xvii.

[A Buzzard was one of the chief dishes in Lieutenant Slicer's valiant dinner, for which see Cartwright's "The Ordinary," ii. 1.]

Cabbage or Cole or Colewort.

Good worts! good cabbage!
MERRY WIVES OF WINDSOR, i. 1, 124.

THE tombstone of the introducer of Cabbage into England is said to exist at Wimborne, probably the Sir Anthony Ashley who was (according to Anthony-à-Wood) a woman-hater. Evelyn, in his "Acetaria" (1699), says, "'Tis scarce an hundred years since we first had Cabbages out of Holland," in which statement he must be mistaken, as Cabbage was commonly eaten all over England before 1633. (*Johnson's Gerard's* "Herbal," p. 313.)

First men ate Coles ere they had corn and flesh to eat; tofore the flood men ate apples, Coles and herbs, as beasts eat grass and herbs. The stalks and leaves thereof grow swifter than stalks and leaves of other herbs; and the overmost crop thereof is called thyme; and the natural virtue of this herb is namely in the crop thereof. The herb breedeth thick blood and troubly and horrible smell. And some Cole is summer Cole, and some is winter Cole. The malice thereof is withdrawn if it be sod or boiled in water, and that water thrown away, and the Cole then sodden in other water with good fatness and savoury. Leaves thereof, bruised and laid to two days, healeth wounds of hounds both new and old, and that wonderly. Cole withstandeth wine and drunkenness, and comforteth the sinews. And the juice thereof helpeth against venom, and also against biting of a wood hound; and serpents flee the smell of Cole sod.

Bartholomew (*Berthelet*), bk. xvii. § 114.

[Gerard in his "Herbal" describes the following sorts of Coleworts: Garden Colewort, curled garden Cole, red Colewort, white Cabbage Cole, red Cabbage Cole, open Cabbage Cole, double Colewort, double crisp or curled Colewort, cauliflower, swollen Colewort (blue and curly), Savoy Cole, curled Savoy Cole, parsley Colewort, and small-cut Colewort; and sea-Colewort (which may be a wild sea-kale), and wild Colewort, grown for its seeds.]

The Colewort being eaten is good for them that have dim eyes, and that are troubled with the shaking palsy. The raw Colewort being eaten before meat doth preserve a man from drunkenness; the reason is yielded, for that there is a natural enmity between it and the vine, which is such, as if it grow near unto it, forthwith the vine perisheth and withereth away; yea, if wine be poured unto it while it is in boiling, it will not be any more boiled, and the colour thereof quite altered. The seed taketh away freckles of the face and sun-burning.

Gerard's "Herbal," bk. ii. ch. xl.

Calf.

The steer, the heifer, and the calf
Are all called neat.

WINTER'S TALE, i. 2, 124.

THE Calf when he is calved hath a certain black spot in the forehead, and witches mean that that speck or whelk exciteth love; but the mother biteth away this speck out of the Calf's forehead, and receiveth him not to her teats, ere the foresaid venom be taken off and done away.

Bartholomew (*Berthelet*), bk. xviii. § 111.

Camel.

Of no more soul nor fitness for the world
Than camels in the war, who have their provand
Only for bearing burdens, and sore blows
For sinking under them.

CORIOLANUS, ii. 1, 266.

CAMELS be beasts that bear charges and burthens, and be mild and soft, and ordained to bear charge and carriage of men. The Camel of Arabia hath two bunches on the back, and the Camel of Bactria hath but one in the back, on the which he beareth his burthen and charge, and another on the breast, and leaneth thereon. And the Camel hateth the horse by kind, and suffereth thirst four days, and stirreth the water with his feet when he drinketh, or else the drink doth him no good. Among four-footed beasts Camels wax bald as men do, and as the ostrich and certain beasts among fowls. Camels have the podagra and the frenzy, and by the podagra their feet be strained, and this evil slayeth them sometime. The Camel is the most hottest beast of kind, and is therefore lean by kind, for the heat draweth off all fatness of the blood, and therefore the Camel is lean. *Bartholomew* (*Berthelet*), bk. xviii. § 19.

THOSE Camels which are conceived by boars are the strongest, and fall not so quickly into the mire as other. It is disdainful and a discontented creature. In the Lake of Asphaltites, wherein all things sink that come in it, many Camels and bulls swim through without danger.

Topsell, "Four-footed Beasts," pp. 72 and 75.

Camomile.

Though the camomile, the more it is trodden on, the faster it grows.

i. KING HENRY IV., ii. 4, 441.

THE oil compounded of the flowers is a remedy against all wearisomeness. *Gerard's* "Herbal," *s.v.*

THOUGH the Camomile, the more it is trodden and pressed down, the more it spreadeth; yet the violet the oftener it is handled and touched the sooner it withereth and decayeth. *Lilly*, "Euphues' Golden Legacy."

THE Camomile shall teach thee patience, which thriveth best, when trodden most upon.

"The More the Merrier" (1608), quoted by Steevens.

Caper.

SIR AND. Faith, I can cut a caper.
SIR TOBY. And I can cut the mutton to 't.
TWELFTH NIGHT, i. 3, 129.

THEY stir up an appetite to meat. They are eaten boiled (the salt first washed off) with oil and vinegar as other salads be, and sometimes are boiled with meat. They be rather a sauce and medicine than a meat.

Gerard's "Herbal," *s.v.*

THEY say that those who eat them daily are in no danger of paralysis. They should not be eaten without coriander. *Hortus Sanitatis*, bk. i. ch. xcvii.

Capon.

Item, A capon, 2*s.* 2*d.*
i. KING HENRY IV., ii. 4, 584.

He steps me to her trencher, and steals her capon's leg.
TWO GENTLEMEN OF VERONA, iv. 4, 10.

THE Capon sitteth on brood upon eggs that be not his own, as it were an hen, and companieth with hens, and eateth with them of their meat, but he feedeth them not; he is fatted with them but he fatteth not them. And sometime his feet are broken to compel him to sit on brood upon eggs. When he is fat, his feet be bound together, and his head hangeth down towards the ground, and is borne by the feet to fairs and to markets. And their brain is better and more profitable than the brains of other fowls. *Bartholomew* (*Berthelet*), bk. xii. § 17.

A CAPON if he be well beaten with nettles will lead its chickens about like a hen, which as they say, he does not for the good of the chickens, but for his own good, that by the warmth of the chickens he may make the poison of the nettle to evaporate. *Hortus Sanitatis*, bk. iii. ch. liv.

ALLECTORIA [or Electorius], is a stone that is found in the maws of Capons, and is like dim crystal. And as witches tell, it is supposed that in battle-fighting, this stone maketh men insuperable, and maketh a man gracious and steadfast, and victor, wise and ready and cunning in plea, and accordeth friends, and quencheth thirst in the mouth.

Bartholomew (*Berthelet*), bk. xvi. § 17.

Caraway.

We will eat a last year's pippin of my own graffing, with a dish of caraways, and so forth.

ii. King Henry IV., v. 3, 3.

[Whether "Caraways" is a kind of apple, or the well-known seeds, the learned commentators on Shakespeare have left undecided. To the many references in *Steevens*' Shakespeare may be added *Dekker's* "Bankrupt's Banquet" and *Heywood's* "Fair Maid of the West," in both which places the seeds are alluded to. Possibly Caraway-seeds were to be eaten with the pippin to correct its crudity, for Gerard says that they are very good for the stomach, help digestion, assuage and dissolve all windiness ("Herbal," *s.v.*).

Sir John Neville at the marriage of his daughter in 1530 provided among a great quantity of other spices "1 pound of Caraways" for one shilling.]

Carbuncle.

A carbuncle entire, as big as thou art,
Were not so rich a jewel.

Coriolanus, i. 4, 55.

A Carbuncle is a precious stone, and shineth as fire, whose shining is not overcome by night. And the kinds thereof be twelve; and is gendered in Lybia among the Troglodytes. Among these twelve manner kinds of carbuncles, those anthracites be the best that have the colour of fire, and be beclipped [enclosed] in a white vein; which have this property—if it be thrown in fire, it is quenched as it were among dead coals, and burneth if water be thrown thereon. And this precious stone is of great price without comparison in respect of other. It is said that it withstandeth graving. And if it be sometime graved and printed with wax, it taketh with him a part of the wax, as it were with biting of a beast.

Bartholomew (*Berthelet*), bk. xvi. § 26.

Carnation. *V.* Gilliflower.

Carp.

Here is a purr of fortune's, sir, or of fortune's cat—but not a musk-cat—that has fallen into the unclean fish-pond of her displeasure, and, as he says, is muddied withal; pray you, sir, use the carp as you may.
ALL'S WELL THAT ENDS WELL, v. 2, 20.

THE Carp is a fish with scales like gold living in lakes or rivers. This fish has much cunning, so that it evades the net. For when it has entered the net, it swims round to look for the opening; and if it cannot find it, it tries to jump over the net so as to get into the open air. Sometimes it seeks a refuge under the net; sometimes it holds sea-weed in its mouth at the bottom of the water, so as to get over the net and escape; sometimes coming with a rush from above, it fixes its head firmly in the mud, so as to escape capture by getting its tail over the net. Its brain is said to grow and diminish as the moon waxes and wanes; and though this holds with all fishes, yet especially so in this one, as among quadrupeds in the wolf and the dog. *Hortus Sanitatis*, bk. iii. ch. xviii.

V. **Fish.**

Cat.

Hang me in a bottle like a cat, and shoot at me.
MUCH ADO ABOUT NOTHING, i. 1, 259.
[*Cf.* Steevens' notes.]

THE Cat falleth on his own feet when he falleth out of high places, and unneath is hurt when he is thrown down off an high place. And when he hath a fair skin, he is as it were proud thereof, and goeth fast about; and when his skin is burnt, then he bideth at home; and is oft for his fair skin taken of the skinner, and flain and slain.

Bartholomew (*Berthelet*), bk. xviii. § 76.

WILD Cats flee from the smoke of rue, and bitter almonds. The dirt of the Tom or of the She-Cat with mustard and vinegar cures baldness.

Hortus Sanitatis, bk. ii. ch. xxv.

It is an unclean and a poisonous animal. It is said to fight against toads, and though it be beaten off by their venomed darts, yet it is not killed.

Hortus Sanitatis, bk. ii. ch. ci.

If dogs chance to find a Cat's skin, they will rub and roll themselves upon it. And they will do so likewise where it is buried; they delight so much of the thing dead, which they hated alive.

Lupton, "A Thousand Notable Things," bk. i. § 77.

Cats are of divers colours, but for the most part grizzled, like to congealed ice, which cometh from the condition of her meat. If the long hairs growing about her mouth be cut away, she loseth her courage. There was in a certain monastery a Cat nourished by the monks, and suddenly the most part of the monks which used to play with the Cat fell sick; whereof the Physicians could find no cause, but some secret poison, and all of them were assured that they never tasted any. At the last a poor labouring man came unto them, affirming that he saw the Abbey-cat playing with a serpent, which the Physicians understanding presently conceived that the serpent had emptied some of her poison upon the Cat, which brought the same to the monks, and they by stroking and handling the Cat were infected therewith; and whereas there remained one difficulty, namely, how it came to pass the Cat herself was not poisoned thereby, it was resolved, that forasmuch as the serpent's poison came from him but in play and sport, and not in malice and wrath, that therefore the venom thereof, being lost in play, neither harmed the Cat at all, nor much endangered the monks; and the very like is observed of mice that will play with serpents. A Cat is much delighted to play with her image in a glass, and if at any time she behold it in water, presently she leapeth down into the water which naturally she doth abhor; but if she be not quickly pulled forth and dried she dieth thereof, because she is impatient of all wet. Those which will keep their Cats indoors, and from hunting birds abroad, must cut off their ears, for they cannot endure to have drops of rain distil into them, and therefore keep themselves in harbour. They cannot abide the savour of ointments, but fall mad thereby. It is most certain that the breath and savour of Cats consume the radical humour and destroy the lungs, and therefore they which keep their Cats with them in their beds have the air corrupted, and fall into several hectics and consumptions. There was a certain company of monks much given to nourish and play with Cats, whereby they were so infected, that within a short space none of them were able either to say, read, pray or sing in all the monastery. And therefore also they are dangerous in the time of pestilence, for they are not only apt to bring home venomous infection, but to poison a man with very looking upon him; wherefore there is in some men a natural dislike and abhorring of Cats. The flesh of Cats can seldom be free from poison, by reason of their daily food, eating rats and mice, wrens and other birds which feed on poison; and, above all, the brain of a Cat is most venomous, by reason whereof memory faileth, and the infected person falleth into a frenzy. But a Cat doth as much harm with her venomous teeth. The hair also of a Cat, being eaten unawares, stoppeth the artery and causeth suffocation. It must needs be an unclean and impure beast that liveth only upon vermin and by ravening, for it is commonly said of a man when he [s]neezeth—that he hath eaten with Cats; likewise, the familiars of witches do most ordinarily appear in the shape of Cats, which is an argument that this beast is dangerous to soul and body.

Topsell, "Four-footed Beasts," pp. 81-3.

Caterpillar.

Caterpillars eat my leaves away.
ii. HENRY VI., iii. 1, 90.

WHEN the rainbow toucheth the tree, no Caterpillars will hang on the leaves. *Lilly*, Epilogue to "Campaspe."

IF you would destroy Caterpillars, do thus: Anoint all the bottom of the tree round about with tar, then get a great sort of ants or pismires, and put them in some bag, and draw the same by a cord unto the tree, and so let it hang there, so that it touch the body of the tree, and the ants letted to go down from the tree by the means of the tar will for want of food eat and destroy all the Caterpillars there, without hurting any of the fruit. This was told me for a very truth.

Lupton, "A Thousand Notable Things," bk. x. § 51.

THE Malshrag [*i.e.*, Caterpillar] is a nesh [soft] worm and full of matter, distinguished with divers colours, shining as a star by night, and hath many colours and foul shape by day. And is not without some pestilential venom, for when he creepeth upon an hot member of a man, he scaldeth the skin, and maketh whelks [*i.e.*, pustules] arise.

Bartholomew (*Berthelet*), bk. xviii. § 47.

V. Vermin and Worm.

SOME Caterpillars are the offspring and breed of dew, as common experience can witness. All Caterpillars are not converted into aurelias [chrysalis], but some of them being gathered and drawn together on a heap (as the vine-fretters), do grow at length to putrefaction, from which sometimes there falleth as it were three blackish eggs, the true and proper mothers and breeders of flies and cantharides. There is not any one sort of Caterpillars, but they are malign, naught and venomous. If you rub a naughty or a rotten tooth with the colewort-Caterpillars, and that often, within a few days following, the tooth will fall out of his own accord. Caterpillars mixed with oil do drive away serpents. *Topsell*, "History of Serpents," pp. 668-70.

Cat o' Mountain.

More pinch-spotted make them
Than pard or cat o' mountain.
TEMPEST, iv. 1, 262.

V. Pard.

IN the Senators' Palace [at Florence] I saw a Cat of the Mountain, not unlike to a dog, with the head of a black colour, and the back like an hedgehog, a light touch whereof gave a very sweet scent to my gloves.

Fynes Moryson, "Itinerary," part i., p. 149.

Cedar.

He shall flourish,
And, like a mountain cedar, reach his branches
To all the plains about him.
KING HENRY VIII., v. 5, 54.

CEDAR is a tree with merry smell, and endureth and abideth long time, and is never destroyed with moth, neither with the tree-worm. Then the Cedar-tree is always green with good smell, and the smell of it driveth away serpents and all manner of venomous worms. And the apple of Cedar hath three manner savours.

Bartholomew (*Berthelet*), bk. xvii. § 23.

Evelyn ("Sylva," bk. ii., ch. iv.) says that chests and presses of Cedar-wood corrupt woollen cloth and furs, but preserve other goods from moths, and, indeed, that the dust and very chips are exitial to moths and worms; that the oil yielded by the wood above all other best preserves books and writings.

Of all trees the Cedar is greatest, and hath the smallest seeds.
Lilly, "Galatea."

The Cedar's juice, whose bitter poison gives
The most strong body unavoided death,
Procures the carcase by its dying force
Void of corruption. *Glapthorne*, "Hollander."

If his malady grow out of ambition, a top of Cedar or an oak-apple is very sovereign with the spirit of hemp-seed.
Brome, "Court-beggar," iii. 1.

Chameleon.

Though the chameleon Love can feed on the air, I am one that am nourished by my victuals.
Two Gentlemen of Verona, ii. 1, 178.

I can add colours to the chameleon.
iii. King Henry VI., iii. 2, 191.

Chameleon is a little beast with divers colours, and his body changeth full soon into divers colours. For it is a fearful beast with little blood, and changeth therefore colours. And is four manner divers [*quadrupes* (*Bartholomew*)]: he hath the face of the eft, and sharp claws and crooked, and the body sharp, and an hard skin as the crocodile. And his sides be even long to the nether parts of his womb as it were a fish; his face is as it were a beast compounded of a swine and of an ape; and his tail is full long and small at the end; and his feet be crooked as it were a little eft; and each of his feet is departed atwin [in two], and the comparison of one foot to another is as in

comparison of the thumb of a man to the other deal of the hand; and each of those two parts is divided in fingers; and his claws be like to the claws of a bird, and all his body is rough and sharp as the body of a Bardan [partan (*Bartholomew*), (? pard or crab)]. His eyes be deep, great and round, and contained with a skin, like to the skin of the body, and that skin covereth the eyes. And he turneth and casteth oft his eyes hither and thither. And changeth his colour when his skin is blown, and his colour is somewhat black with black speckles therein. And this diversity is in all his body, and namely in the eyes, and also in the tail, and is full heavy in moving and foul of colour in his death, and what is in his body is but of little flesh; and hath little blood, but in the head, and in the end of the tail, where he hath little blood and also in the heart, and in the veins that come therefrom; and also hath blood about the eyes, though it be right little. His most might and strength is against the kind of goshawks; for he draweth them, and they fly to him, and he taketh them wilfully to other beasts to be devoured. If his head and his throat be set afire with oaken wood, it maketh both rain and thunder. In sickness he feigneth himself soft and mild, though he be cruel. And it is said, that the Chameleon liveth only by air, and the mole by earth, and the herring by water, and the cricket [salamander (*Bartholomew*)] by fire.

Bartholomew (*Berthelet*), bk. xviii. § 21.

If the Chameleon at any time see a serpent taking the air, and sunning himself under some green tree, he climbeth up into that tree, and settleth himself directly over the serpent, then out of his mouth he casteth a thread like a spider, at the end whereof hangeth a drop of poison as bright as any pearl, which lighting upon the serpent killeth it immediately. The right claw of the fore-feet, bound to the left arm with the skin of his cheeks, is good against robberies and terrors of the night, and the right pap against all fears. If the left foot be scorched in a furnace with the herb Chameleon, and afterward putting a little ointment to it, and made into little pasties, so being carried about in a wooden box, it maketh the party to go invisible. Likewise the liver dissolveth amorous enchantments. The entrails and dung of this beast washed in the urine of an ape, and hung up at our enemies' gates, causeth reconciliation. With the tail they bring serpents asleep, and stay the flowing of the floods and waters; the same mingled with cedar and myrrh, bound to two rods of palm, and struck upon water, causeth all things that are contained in the same water to appear.

Topsell, "History of Serpents," pp. 675-6.

Cherry.

KING HENRY VIII., v. 1, 169.

[Gerard ("Herbal," *s.v.*) reckons up the following sorts of Cherries: English, Flanders, Spanish, Gascon (late-ripe), Chester, double-flowered, barren double-flowered, bird's Cherry or black grape Cherry, another bird's Cherry, common black Cherry, dwarf Cherry, greater and lesser heart Cherry (or Luke Ward's Cherry and the Naples Cherry), large black Cherry, agriot, large-fruited dwarf, and dun-coloured Cherry, besides many other unnamed sorts. Later, he speaks of Kentish Cherries and Morella Cherries, but he has no good word to say of any of them.

Evelyn ("Sylva," ch. xx.) says that Cherries were said to have been brought into Kent out of Flanders by Henry VIII.]

As many several change of faces
As I have seen carved upon a Cherry-stone.

Webster's "Devil's Law Case," iii. 4.

Chestnut.

Do you tell me of a woman's tongue,
That gives not half so great a blow to hear
As will a chestnut in a farmer's fire?

TAMING OF THE SHREW, i. 2, 208.

A CASTEIN tempered with a little honey healeth at the best biting of a wood hound, or man's biting. Also the rinds and leaves burnt and made to powder, tempered with vinegar and laid to a young man's head in a plaster-wise, maketh hair increase, and keepeth hair from falling.

Bartholomew (*Berthelet*), bk. xvii. § 88.

THE Chestnut is next the oak one of the most sought after by the carpenter and joiner. It hath formerly built a good part of our ancient houses in the City of London, as does yet appear. If water touch the roots of the growing trees, it spoils both fruit and timber. The beams made of Chestnut-tree have this property, that, being somewhat brittle, they give warning, and premonish the danger by a certain crackling which it makes.

Evelyn's "Sylva," bk. i. ch. viii.

MAIDS, if you look to roast your Chestnuts well,
Observe first with a knife to wound the shell;
If with unbroken skin it touch the fire,
'Twill break in pieces, and with noise retire.

Heywood's "Anna and Phillis," emb. 33.

Chick, Chicken.

TROILUS AND CRESSIDA, i. 2, 147.

A WALNUT put fast in a Chicken, that it fall not out in the roasting thereof; it makes that the same Chicken will be the sooner roasted.

Lupton, "A Thousand Notable Things," bk. v. § 32.

Chrysolite.

If heaven would make me such another world
Of one entire and perfect chrysolite,
I'ld not have sold her for it.

OTHELLO, v. 2, 143.

CHRYSOLITE is a little stone of Ethiopia, shining as gold, and sprinkling as fire, and is like to the sea in colour, and somewhat green. If it be set in gold and

borne on the left arm, it feareth fiends and chaseth them away, and it helpeth night-frays and dreads; and abateth an evil that hight melancholy, or doth it away; and comforteth the in-wit. One manner of Chrysolite is deemed golden by day, and fiery by night. And another manner kind is coloured as gold, and is right fair in sight in the morrow tide; and then as the day passeth his colour waxeth dim. And this stone taketh most soonest heat; for if it be set by the fire, anon it waxeth on aflame.

Bartholomew (*Berthelet*), bk. xvi. § 29.

According to the *Hortus Sanitatis* (bk. iv. § 38), Chrysolite drives away demons and the worst melancholy fears if pierced, and the hole filled up with ass's bristles, and the stone bound on the left arm. And some say that it drives away folly, and brings wisdom.

CHRYSOLITE, the purer the sooner stained.

"Euphues' Golden Legacy."

Civet.

A' rubs himself with civet.

MUCH ADO ABOUT NOTHING, iii. 2, 45

Civet is of a baser birth than tar, the very uncleanly flux of a cat.

AS YOU LIKE IT, iii. 2, 69.

THIS is Civet, this comes from the cat's tail, this perfumes your ladies, this drug is precious and dear.

Sharpham, "The Fleire."

I vow to poison your musk-cats, if their Civet excrement do but once play with my nose.

Dekker's "Gull's Hornbook," bk. ii.

HE wears Civet,
And when it was ask'd him where he had that musk,
He said all his kindred smelt so.

"Soliman and Perseda," i.

(CIVET as an ingredient of a pomander.) "Lingua," iv. 3.

MUSK-CAT, I'ld make your Civet worship stink
First in your perfumed buff.

Thomas Rawlins, "The Rebellion," ii. 1.

THIS beast is a very clean beast, and therefore the place where it lieth must be swept every day and the vessels clean washed. The Civet or liquor running out doth go back again if any vessel be put to receive it, except it be a silver spoon or porringer. This Civet is nothing else but the sweat of the beast under the ribs, fore-legs, neck and tail.

Topsell, "Four-footed Beasts," p. 586.

Cloves.

BIRON. A lemon.
LONG. Stuck with cloves.

LOVE'S LABOUR'S LOST, v. 2, 653.

HERE'S New Year's gift has an orange and rosemary, but not a Clove to stick in 't.

Ben Jonson, "Masque of Christmas."

WINE will be pleasant in taste and in savour and colour; it will much please thee, if an orange or a lemon (stuck round about with Cloves) be hanged within the vessel that it touch not the wine. And so the wine will be preserved from foistiness and evil savour.

Lupton, "A Thousand Notable Things," bk. ii. § 40.

HE walks most commonly with a Clove or pick-tooth in his mouth. *Ben Jonson*, "Cynthia's Revels," ii. 3.

IN the goose-market numbers of freshmen stuck here and there with a graduate, like Cloves with great heads in a gammon of bacon. *Webster's* "Northward Ho!" i. 1.

THAT Westphalian gammon Clove-stuck face.

Marston's "Scourge of Villainy," Satire vii., line 114.

SOME be feigned with powder of good Cloves meddled with vinegar and wine with good smell, and be unneath known. But these that be feigned may not be kept passing twenty days. Good Cloves comfort the brain and the virtue of feeling, and help also against indignation and ache of the stomach. *Bartholomew* (*Berthelet*), bk. xvii. § 79.

Cock.

TAMING OF THE SHREW, ii. 1, 227-8.

COCK'S flesh raw, and laid hot upon the biting of a serpent, doth away the venom. And to the same his brain is good, taken in drink. And if a man be [a]nointed with his grease, or with his juice, he shall be sure from panthers and lions. And if the bones of a Cock or of an hen be meddled with gold when it is molten, they destroy and waste the gold. And so hen's bones be venomous to gold, and that is wonder. When he hath the mastery [over his adversaries] he singeth anon; and ere he singeth, he beateth himself with his wings to make him the more able to sing. And he useth far in the night to sing most clearly, and to sing strongly. And about the morrow-tide he shapeth light voice and song. The Cock beareth a red comb on his head instead of a crown; which being lost he loseth his hardiness, and is more slow and coward to assail his adversary. And he setteth next to him on the roost the hen that is most fat and tender, and loveth her best; in the morrow-tide, when he flieth to get his meat, first he layeth his side to her side; and he fighteth for her specially as though he were jealous. And he breedeth a precious stone called allectricium, [or allectoria, *v.* Capon] like to the stone that hight chalcedony; and the Cock beareth that stone, and by cause of that stone (as some men trow) the lion dreadeth and abhorreth [him], and specially if the Cock be white. For the lion dreadeth the white Cock. Also the Cock dreadeth the eagle and the goshawk, which take their prey on the ground. And the Cock is right sharp of sight, and therefore he looketh downward with the one eye to seek his meat, and upward into the air with the other eye to beware of coming of the eagle and of the goshawk. Also a right aged Cock layeth eggs in his last end, and the eggs are small and full round, and as they were wan or yellow. And if any venomous worm sitteth on brood on them in the canicular days, of them be bred and grow cockatrices. *Bartholomew* (*Berthelet*), bk. xii. § 16.

It is to be marvelled at, that a Cock or Cockerel which doth not fear a serpent or a dragon is so afraid of the shadow of a glead, when he is flying, that suddenly he seeks a place of refuge, and hides himself.

Lupton, "A Thousand Notable Things," bk. i. § 24.

If the blood of a Cock be dried, and made in powder, and mixed in wine, wherein there is water, it makes the water swim above. This was the relation of a learned monk. *Ibid.* bk. vi. § 6.

If any man wishes that a Cock should not crow, let him anoint its head and brow with oil.

Albertus Magnus, "Of the Wonders of the World."

In the beginning of the night God causeth all the gates of heaven to be shut, and the Angels stay at them in silence, and sendeth evil spirits into the world, which hurt all they meet; but after midnight they are commanded to open the same. This command and call is heard of the Cocks, and therefore they clap their wings and crow to awaken men; and then the evil spirits lose their power of hurting.

Purchas, "Pilgrims," p. 194 (ed. 1616); *cf.* Hamlet, i. 1, 147-155.

Cockatrice. *V.* Basilisk.

Cockle.

Hamlet, iv. 5, 25.

The flesh of river Cockles, whether raw or cooked, resist the stings of scorpions. *Hortus Sanitatis*, bk. iii. ch. xxiii.

Colchester oysters and your Selsey Cockles.

Ben Jonson, "The Fox," ii. 1.

Have our Cockles boiled in silver shells.

Ben Jonson, "The Alchemist," iv. 1.

You may eat the cramm'd Cockle.

Middleton, "A Game at Chess," v. 3, 70.

Cockle (plant).

Love's Labour's Lost, iv. 3, 383.

Ray is a certain herb; poets call this herb ungracious Cockle or weed; and it groweth among wheat in corrupt time and dry. And ray hath a sharp strength and working, and somedeal venomous, and maketh men drunk, and disturbeth the wit, and grieveth the head, and changeth savour of bread and infecteth bread that it is meddled with, and grieveth full soon, and slayeth sometime if it be eaten in great quantity. *Bartholomew* (*Berthelet*), bk. xvii. § 194.

V. Darnel.

Codling.

Not yet old enough for a man, nor young enough for a boy; as a squash is before 'tis a peascod, or a codling when 'tis almost an apple.
TWELFTH NIGHT, i. 5, 167.

[So in *Ben Jonson's* "Alchemist," and *Brome's* "Mad Couple," i. 1.]

V. Apple.

Columbine.

There's fennel for you, and columbines.
HAMLET, v. 4, 180.

[To the notes in *Steevens'* Shakespeare on this passage may be added that the Columbine was also called *Herba Leonis*, or "the herb wherein the lion doth delight," and that it was "used especially to deck the gardens of the curious, garlands and houses" (*Gerard's* "Herbal," *s.v.*).

Minsheu (Dictionary, *s.v.*) translates Columbine into the Latin *Aquilegia* ("because in its flowers there is some likeness to the eagle"), and *Chelidonia* (*i.e.*, celandine), which is so called, "for it springeth or bloometh in the coming of swallows." "By the juice of celandine swallows' eyes turneth again to the first state, if they be hurt or put out" (*Bartholomew,* bk. xvii. § 46). So *Lupton* ("A Thousand Notable Things," bk. iii. § 89): "The eyes of young swallows being in the nest, pricked with a needle or a pin, and so made blind, within four or five days after, they will see again; which is very true, for I have proved it. But how they recover their sight I know not. But divers write, if their eyes be hurt, the old swallows restore their sight again with the juice of celandine." And the same author states: "Celandine with the heart of a wont or a mould-warp [*i.e.*, mole], laid under the head of one that is grievously sick, if he be in danger of death, immediately he will cry with a loud voice, or sing: if not, he will weep" (bk. ii. § 4).

Most probably Minsheu translated "Columbine" wrongly, but the virtues of celandine are worthy of record.]

Cony.

They will out of their burrows like conies after rain.
CORIOLANUS, iv. 5, 226-7

CONIES be called small hares and feeble, and they dig the earth with their claws, and make them bowers and dens under the earth, and dwell therein, and bring forth many rabbits, and multiply right much. And rabbits be so loved in the Balearic Isles that those rabbits without mothers be taken and eaten of the men of the country, though the guts be unneath cleansed. As many dens as be in the increasing [excrement (*Bartholomew*)] of the Conies, so many years they have of age. In [that part of] the body be so many holes as the Conies have years. It is said that they have both sexes, male and female. And is a profitable beast both to meat and to clothing, and to many manner medicines. *Bartholomew* (*Berthelet*), bk. xviii. § 68.

BY night he devours vine-shoots and fruits, but in the morning he enters his den, and makes the opening of it level with the soil by dust from within, lest men coming past by day should find out his dwelling.

Hortus Sanitatis, bk. ii. ch. xlv.

Conger.

Eats conger and fennel.
ii. KING HENRY IV., ii. 4, 229.

CONGER is a sea-fish, as long as a lamprey, but much larger in the body. When the wind blows strongly it grows fat, and its flesh is most sweet to eat. It is an enemy to lampreys and other fish, yet it is strong, so that it can tear a polypus by the strength of its teeth. The Conger and the lamprey hate one another, and bite each other's tails. *Hortus Sanitatis*, bk. iii. ch. xxiv.

THE Conger hath many wiles, and is witty and wily of getting of meat, for when he seeth meat on a hook, he dreadeth the hook, and biteth not the bait, but holdeth the hook with his fins, and letteth it not pass till he have gnawn the meat. *Bartholomew* (*Berthelet*), bk. xiii. § 29.

FENNEL was commonly eaten with Conger.

Ben Jonson, "Bartholomew Fair," and "Philaster."

Copper.

i. KING HENRY IV., iii. 3, 162.

V. **Brass.**

COPPER is lately not found, but restored again to light. Strangers have most commonly the governance of our mines.

Holinshed, "Description of England," p. 238.

Coral.

Full fathom five thy father lies;
Of his bones are coral made.

TEMPEST, i. 2, 397.

CORAL is gendered in the Red Sea, and is a tree as long as it is covered with water; but anon as it is drawn out of water and touched with air, it turneth into stone. Witches tell that this stone withstandeth lightning. His might and virtue is wonderful, for it putteth off lightning, whirlwind, tempest and storms from ships and houses that it is in. And it is double white and red, and is never found passing half a foot long. And the red helpeth against the fiend's guile and scorn, and against divers wondrous doing, and multiplieth fruit, and speedeth beginning and ending of causes and of needs.

Bartholomew (*Berthelet*), bk xv. § 33.

[*Sir Thomas Browne* ("Vulgar Errors," bk. ii. ch. v.), doubts whether Coral be soft under water, and adds that "a gentleman caused a man to go down into the sea no less than a *hundred* fathom to see if it were so." Truly there were divers in those days! Further (bk. v. ch. xxiii.) he says: "Though Coral doth properly preserve and fasten the teeth in men, yet it is used in children to make an easier passage for them, and for that intent is worn about their necks."].

Cow.

The breese upon her, like a cow in June
Hoists sails and flies.
ANTONY AND CLEOPATRA, iii. 10, 14.

WHEN the kine do oft calve and have many calves, it is a token as men mean that in winter shall be much rain. And when they have sore feet, it is medicine therefor to anoint them between the horns with oil and pitch and other medicines. And have the gout and die of that evil; and the token thereof is when they bear down their ears, and eat not. And when she is stung with a great fly, then she raiseth up her tail in a wonder wise, and startleth as she were wood about fields and plains.

Bartholomew (Berthelet), bk. xviii. § 109.

THE hoofs of the fore-feet of a Cow dried, and made in fine powder, increaseth milk in nurses if they eat it in their pottage, or use it in their drink; and being cast upon burning coals, the smoke thereof doth kill mice, or at the least doth drive them away.

Lupton, "A Thousand Notable Things," bk. i. § 4.

Cowslip.

Freckled cowslip.
KING HENRY V., v. 2, 49.

COWSLIP, because the cow licketh this flower up with her lips. *Minsheu's* Dictionary, *s.v.*

COWSLIPS [or] two-in-a-hose. *Gerard's* "Herbal," *s.v.*

Crab.

If like a crab you could go backward.
HAMLET, ii. 2, 206.

GREAT cold grieveth them [*i.e.*, fish] sore, and namely them that have stones in their heads as Crabs and other such. For the stone in the head runneth and freezeth, and such a fish dieth soon. Also the Crab is enemy to the oyster, for he liveth by fish thereof with a wonderful wit. For because that ye [? he] may not open the hard shell of the oyster, he spieth and awaiteth when the oyster openeth, and then the Crab (that lieth in await) taketh a little stone, and putteth between the shells, that the oyster may not close himself. And when the closing is so let, the Crab eateth and gnaweth the fish of the oyster.

Bartholomew (*Berthelet*), bk. xiii. § 29.

THE Crab goes backward, and has never known how to follow his nose. When he grows old, two stones of a white colour mixed with red are found in his head, which are said to be of such virtue that, given in drink, they heal punctures of the heart. There are some little Crabs on the coast of Judea which are called soldiers, because they run so fast and cannot be caught. And if one of them be cut in half, there is no flesh or superfluity at all to be found in its body, because they take no food.

Hortus Sanitatis, bk. iv. ch. xvi.

[IN the Moluccas shipwrecked men were forced to build a fort] to defend themselves from certain Crabs of exceeding greatness, and in as great numbers, and of such force, that whosoever they got under their claws it cost him his life. *Purchas'*, "Pilgrims," p. 504, ed. 1616.

CRABS here with us have a sympathy with the moon, and are fullest with her fulness. In India there is a contrary antipathy, for at full moon they are emptiest.

Ibid., p. 505.

Crab (*i.e.*, Crab-apple).

And sometimes lurk I in a gossip's bowl
In very likeness of a roasted crab.

MIDSUMMER NIGHT'S DREAM, ii. 1, 48.

[Lamb's wool was made of cultivated apples, not of crabs. *Cf.* *Gerard's* "Herbal," *s.v.* Apple.]

A CUP of ale had in his hand, and a Crab lay in the fire.

"Gammer Gurton's Needle."

Cricket.

As merry as crickets.

i. KING HENRY IV., ii. 400.

[CRICKET] is a little beast, feeble and mightless and thievish and venomous with pricks and pikes. This beast goeth backward, and saweth and diggeth the earth, and worketh by night; and is hunted with an ant tied with an hair, and thrown into his den: and the powder [dust] is first blown away, lest the ant hide herself therein, and so he is drawn to love of the ant. *Bartholomew* (*Berthelet*), bk. xviii. § 58.

Crocodile.

Your serpent of Egypt is bred now of your mud by the operation of the sun: so is your crocodile.

ANTONY AND CLEOPATRA, ii. 7, 29.

If that the earth could teem with woman's tears
Each drop she falls would prove a crocodile.

OTHELLO, iv. 1, 256.

The mournful crocodile
With sorrow snares relenting passengers.

ii. KING HENRY VI., iii. 1, 226.

O LAND Crocodiles,
Made of Egyptian slime, accursed women!

Massinger, "The Renegado," iii. 1.

THE Crocodile is a serpent that from a small egg, grows in short time to a mighty length and bigness; he is bold over those that fly him, but fearful of them that pursue

him; the four winter months, November, December, January and February, he eats not at all; he hath no tongue, but teeth sharp and long; neither in feeding doth he move his lower jaw.

Thomas Heywood, "London's Peaceable Estate."

Crocodile is nigh twenty cubits long, and his skin is hard that recketh not though he be strongly beaten on the back with stones. And a certain fish, having a crest like to a saw, rendeth his tender womb, and slayeth him. And it is said that among beasts only the Crocodile moveth the over jaw. Among beasts of the land he is tongueless, and his biting is venomous; his teeth be horrible and strongly shapen as a comb or a saw, and no beast that cometh of so little beginning waxeth so great, and is a beast nourished in great gluttony, and eateth right much. And so when

he is full, he lieth by the brink or by the cliff, and bloweth for fullness; and then there cometh a little bird, which is called king of fowls among the Italians, and this bird flyeth tofore his mouth, and sometime he putteth the bird off, and at the last he openeth his mouth to the bird, and suffereth him enter. And this bird claweth him first with claws softly, and maketh him have a manner liking in clawing, and falleth anon asleep, and when this bird knoweth and perceiveth that this beast sleepeth, anon he descendeth into his womb, and forthwith sticketh him as it were with a dart, and biteth him full grievously and full sore. The Crocodile is right nesh and full tender in the womb, and for that cause he is soon overcome of such fishes, which have sharp pricks and crests growing on their backs on high. This grim and most horrible beast followeth and pursueth them that fly, and is dreadful to them; and he fleeth serpents, and hath dim eyes while he is in the water, and seeth too sharply when he is out of water. And he waxeth more all the time that he is alive. If the Crocodile findeth a man by the brim of the water, or by the cliff, he slayeth him if he may, and then he weepeth upon him, and swalloweth him at the last. And of his dirt is made an ointment, and with that ointment women anoint their own faces. And so old women and rivelled [wrinkled] seem young wenches for a time. And the Crocodile eateth gladly good herbs and grass, among whom lurketh a little serpent, and is enemy to the Crocodile, and hideth him privily in the grass, and wrappeth himself therein, and so while the Crocodile eateth grass, he swalloweth this serpent, and this serpent entereth into his womb, and allto [quite] rendeth his guts, and slayeth him, and cometh out harmless. The same worm lieth in await on the Crocodile when he sleepeth, and then wrappeth himself in fen [*i.e.* mud], and entereth in between his teeth, and cometh into his body. The Crocodile lieth in await on certain small birds that breed among the grass of the River Nile, the which birds fly into the womb of the Crocodile for heat of the sun, and eateth the worms of his womb; and so that fierce beast is cleansed and purified of worms. And so dwelleth in land by day, and in water by night; for the water is hotter by night than by day, for the water holdeth the sunbeams, and be moved, and so the water is hot.

Bartholomew (*Berthelet*), bk. xviii. § 33.

Of late years, there hath been brought into England the cases or skins of such Crocodiles, to be seen,—and much money given for the sight thereof, the policy of strangers; laugh at our folly, either that we are too wealthy, or else that we know not how to bestow our money.

Batman's addition to *Bartholomew*, bk. xviii. § 33.

His nature is ever when he would have his prey to cry and sob like a Christian body, to provoke them to come to him, and then he snatcheth at them; and thereupon came this proverb that is applied to women when they weep, *Lachrymæ crocodili*, the meaning whereof is, that as the Crocodile when he crieth, goeth then about most to deceive, so doth a woman most commonly when she weepeth.

Master John Hawkins' "Second Voyage"
apud Hakluyt, p. 534 (ed. 1598).

The Crocodile is a great worm, abiding near the rivers sides. The Crocodile of the earth is afraid of saffron, and therefore the country-people, to defend their hives of bees and honey from them, strew upon the places saffron. It is doubtful whether it hath any place of excrement except the mouth. They do not cast their skins as other serpents do. After the egg is laid by the Crocodile, many times there is a cruel stinging scorpion which cometh out thereof, and woundeth the Crocodile that laid it. The Crocodile is a fearful serpent, abhorring all manner of noise, especially from the strained voice of a man. The Crocodile runneth away from a man if he wink with his left eye, and look steadfastly upon him with his right eye. Because he knoweth that he is not able to overtake a man in his course or chase, he taketh a great deal of water in his mouth, and casteth it in the path-ways, so that when they endeavour to run from the Crocodile, they fall down in the slippery path. There is an amity and natural concord betwixt swine and Crocodiles. If but a feather of the ibis come upon the Crocodile by chance, or by direction of a man's hand, it maketh it immoveable and cannot stir. There is a kind of thorny wild bean growing in Egypt, this is a great terror to the Crocodile, for he is in great dread of his eyes, and therefore all the people bear them in their hands when they travel. When they go to the land to forage and seek after a prey, they cannot return back again, but by the same footsteps of their own which they left imprinted in the sand [and so they may be caught in a trench made in their path]. The Indians have a kind of Crocodile in Ganges, which hath a horn growing out of his nose like a rhinocerot. The blood of a Crocodile is thought to cure the bitings of any serpent. The skin both of the land and water Crocodile dried into powder, and the same powder with vinegar or oil laid upon a part or member of the body to be seared, cut off, or lanced, taketh away all sense and feeling of pain from the instrument in the action. The poison of the Crocodile worketh by cold air and light, and therefore by the want of both is to be cured.

Topsell, "History of Serpents," pp. 683-92.

Crow.

Crows are fatted with the murrion flock.
MIDSUMMER NIGHT'S DREAM, ii. 1, 97.

THE Crow is a bird of long life. And diviners tell that she taketh heed of spyings and awaitings, and teacheth and sheweth ways, and warneth what shall fall. But it is full unlawful to believe, that God showeth his privy council to Crows. Among many divinations, diviners mean that Crows token rain with greding and crying. And is a jangling bird and unmild, and grievous to men there they dwell. And eateth unclean meats and venomous, and liveth right long. In age their feathers wax white; but in flesh within, the longer they live, the more black they be. Crows rule and lead storks, and come about them as it were in routs, and fly about the storks, and defend them. and fight against other birds and fowls that hate storks. And take upon them the battle of other birds upon their own peril. And an open proof thereof is,—for in that time that the storks pass out of the country Crows be not seen in places where they were wont to be; and also for they come again with sore wounds, and with voice of blood that is well known, and with other signs and tokens, and show that they have been in strong fighting. And the mildness of the bird is wonderful. For when father and mother in age be both naked and bare of feathers, then the young Crows hide and cover them with their feathers and gather meat and feed them. And sometime when the father and mother wax old and feeble, then the young Crows underset them and rear them up with their wings, and comfort them to use to fly, to bring the members that be diseased into state again.

Bartholomew (*Berthelet*), bk. xii. § 9.

IN the solstice the Crow is seized with disease; it feeds freely on nuts. It lies in wait for the eggs of the dove, to break them and suck them.

Hortus Sanitatis, bk. iii., ch. xxxiii.

IF a Crow chance to eat of the rest of the flesh whereof a wolf hath eaten before: the same Crow will die soon after.

Lupton, "A Thousand Notable Things," bk. vi. § 49.

Crow-flowers.

HAMLET, iv. 7, 170.

BESIDES these kinds of Pinks before described, there is a certain other kind either of the gilly-flowers or else of the Sweet Williams, altogether and every where wild. I do hold it for a degenerate kind of wild gilly-flower. These grow all about in meadows and pastures and darkish places. They begin to flower in May and end in June. The Crow-flower is called wild Williams, marsh gilly-flowers, and cuckoo gilly-flowers. These are not used in medicine or in nourishment; but they serve for garlands and crowns, and to deck up gardens. *Gerard's* "Herbal," *s.v.*

Crown imperial.

WINTER'S TALE, iv. 4, 126.

THIS plant hath been brought from Constantinople amongst other bulbous roots, and made denizens in our London gardens, whereof I have great plenty.

Gerard's "Herbal," *s.v.*

Crystal.

LOVE'S LABOUR'S LOST, ii. 1, 243.

CRYSTAL is a bright stone and clear with watery colour. Men trow that it is of snow or ice made hard in space of many years. This stone set in the sun taketh fire, insomuch if dry tow be put thereto, it setteth the tow on fire. *Bartholomew* (*Berthelet*), bk. xvi. §31.

Cuckoo.

MIDSUMMER NIGHT'S DREAM, iii. 1, 134.

THE Cuckoo is a dishonest bird, and is very slow, and does not stay in a place. In winter it is said to lose its feathers; and it enters a hole in the earth or hollow trees; there in the summer it lays up that on which it lives in the winter. They have their own time of coming, and are borne upon the wings of kites, because of their short and small flight, lest they be tired in the long tracts of air and die. From their spittle grasshoppers are produced. In the winter it lies languishing and unfeathered, and looks like an owl. *Hortus Sanitatis*, bk. iii. ch. xxxix.

IF you mark where your right foot doth stand at the first time that you do hear the Cuckoo, and then grave or take up the earth under the same,—wheresover the same is sprinkled about, there will no fleas breed. And I know it hath proved true.

Lupton, "A Thousand Notable Things," bk. iii. § 47.

WHEN you first see the Cuckoo, mark well where your right foot doth stand; for you shall find there an hair, which, if it be black, it signifies that you shall have very evil luck all that year following. If it be white, then it signifies very good luck; but if it be grey, then indifferent. It is certain such a hair hath been found accordingly, but what event did follow thereof I am yet uncertain. But this was affirmed unto me for a very truth. It was also credibly reported to me, that the like hair will be found under the right foot at the first seeing of the swallow, after they are come at the spring-time; so that you look after the said swallow, as long as you can see her.

Ibid., bk. x. § 80.

Cuckoo-bud.

When daisies pied and violets blue
  And lady smocks all silver white
And cuckoo-buds of yellow hue
  Do paint the meadows with delight.

LOVE'S LABOUR'S LOST, v. 2, 906.

[If the learned Steevens in writing his note on this passage had noticed that Shakespeare draws a special distinction between the colours of "lady-smocks" and "Cuckoo-buds," he would not have suggested that Shakespeare might not have been sufficiently acquainted with botany to be aware that lady-smocks are also called Cuckoo-*flowers* (which latter word occurs "Lear," iv. 4, 4). "Cuckoo-bud" may be the *Ranunculus bulbosus*, which

Gerard calls the Round-rooted Crow's-foot, or it may be the "Cuckoo-bread," "Cuckoo-brood," or as Gerard calls it, "Cuckoo-meat," *i.e.*, the *Oxalis acetosella*. Gerard describes a variety of this plant with yellow flowers, and says that it has its name "because either the Cuckoo feedeth thereon, or by reason when it springeth forth and flowereth the Cuckoo singeth most."]

Cuckoo-flower.

[See above and *Lady-smock*.]

Currants.

WINTER'S TALE, iv. 3, 40.

[Gerard only casually alludes to the Currant-bush which now grows in England, of which, however, Johnson, in his appendix to Gerard's "Herbal" gives a full description. These currants, therefore, will be currants of Zante or Cephalonia, as Fynes Moryson calls them.]

THE black Currants are used in sauces, and so are the leaves also by many. *Parkinson's* "Herbal," *s.v.*

Cuttle.

An you play the saucy cuttle with me.
ii. KING HENRY IV., ii. 4, 139.

CUTTLE-FISH is a kind of sea-fish, with a pointed snout, with which they pierce and sink ships in the Atlantic Ocean.
Minsheu's Dictionary, *s.v.*

ITS ink is so strong that when thrown on a lamp, men seem to be Ethiopians. It conceives by the mouth like a viper. *Hortus Sanitatis*, bk. iv. ch. lxxxi.

Cypress.

Cypress chests.
TAMING OF THE SHREW, ii. 1, 353.

THIS Cypress-tree is formable and necessary to edifying and building of towers and temples, and for other great and pompous edifices. And for because it may not rot, it faileth never, but abideth and dureth and lasteth always in the first estate and condition; and hath a right good savour, and most sweetest smelling.

Bartholomew (*Berthelet*), bk. xvii. § 24.

CYPRESS groweth in divers places of England where it hath been planted, as at Sion, a place near London, sometime a house of Nuns. *Gerard's* "Herbal," *s.v.*

THE leaves of Cyprus do make the hair red.

Gerard's "Herbal," *s.v.* "Privet."

[Cyprus ("Winter's Tale," iv. 3, 221), *i.e.*, lawn or crape, was so called from the island, whence it first was brought to England.]

Dace.

If the young dace be a bait for the old pike.

ii. KING HENRY IV., iii. 2, 356.

[*Minsheu* (Dictionary, *s.v.*) gives *apua* for the Latin of *Dace*, and *Cooper* ("Thesaurus," *s.v.*) explains *apua*—or, as he writes it, *aphya*—as a "fish having [its] beginning of abundance of rain."]

Daffodils.

Daffodils,
That come before the swallow dares, and take
The winds of March with beauty.

WINTER'S TALE, iv. 4, 118.

[Gerard describes fourteen kinds of Daffodils, to which Johnson adds eighteen more.]

A CATAPLASM made of the root of Daffodil, honey and oatmeal draws forth spills, shivers, arrow-heads, and thorns, and whatsoever stick within the body.

Holland's Pliny, bk. xxi. ch. xix.

Daisy.

Daisies pied.

LOVE'S LABOUR'S LOST, v. 2, 904.

DAISY or Ox-eye: The young roots are frequently eaten by the Spaniards and Italians all the spring till June.

Evelyn's "Acetaria," § 22.

[*Gerard* describes six different Daisies, and states further that the juice of the leaves and root given to little dogs keepeth them from growing great ("Herbal," *s.v.*).]

Damson.

My wife desired some damsons.
ii. KING HENRY VI., ii. 1, 101.

OF the plum-tree is many manner of kinds; but the Damascene is the best, that cometh out of Damask; only of this tree droppeth and cometh glue and fast gum, physicians say that it is profitable to medicine, and for to make ink for writers' use.

Bartholomew (*Berthelet*), bk. xvii. § 125.

Darnel.

Darnel and all the idle weeds that grow
In our sustaining corn.
KING LEAR, iv. 4, 5.

AMONG the hurtful weeds, Darnel is the first. They grow in fields among wheat and barley of the corrupt and bad seed. They spring and flourish with the corn. The new bread wherein Darnel is eaten, hot, causeth drunkenness; in like manner doth beer or ale wherein the seed is fallen, or put into the malt. Darnel hurteth the eyes, and maketh them dim, if it happen in corn either for bread or drink. *Gerard's* "Herbal," *s.v.*

Date.

Your date is better in your pie and your porridge than in your cheek.
ALL'S WELL THAT ENDS WELL, i. 1, 173.

THERE is made hereof both by the cunning confectioners and cooks divers excellent cordial, comfortable and nourishing medicines, and that procure lust of the body very mightily. The ashes of the Date-stones heal falling away of the hair of the eye-lids, being applied together with spikenard. *Gerard's* "Herbal," *s.v.*

Daw.

I am no wiser than a daw.
i. KING HENRY VI., ii. 4, 18.

THE Daw fights with the owl, because the owl has but weak sight by day; for this reason the Daw carries off the owl's eggs and eats them. Its flesh causes itching in the head, for itself loves to be scratched on the head.

It is said to go mad often; so that it often hangs itself in the forked branches of trees.

Hortus Sanitatis, bk. ii. chs. lxxx. and lv.

Dead Men's Fingers.

Long purples
That liberal shepherds give a grosser name,
But our cold maids do dead men's fingers call them.
HAMLET, iv. 7, 171.

[Dead Men's Fingers is the *Orchis mascula*, or, as Gerard calls it, Satyrion Royal or finger orchis; the plant has this name from the shape and colour of the root.]

Deer.

LOVE'S LABOUR'S LOST, iv. 2, 3 *et seq.*

THE most excellent of all animals.

Minsheu's Dictionary, *s.v.*

Cf. Hart.

IN the blood of these kind of Deer [Fallow-Deer] are not strings or fibres, wherefore it doth not congeal as other doth, and this is assigned to be one cause of their fearful nature; they are also said to have no gall. Their blood doth increase above measure melancholy. The dung or fime of this beast, mingled with oil of myrtles, increaseth hair, and amendeth those which are corrupt. Some of the late writers do prescribe the fat of a mole, of a Deer, and of a bear, mingled together to rub the head withal for increase of memory.

Topsell, "Four-footed Beasts," p. 90.

Dew-berry.

MIDSUMMER NIGHT'S DREAM, iii. 1, 169.
Also in *Marlow's* "Dido, Queen of Carthage," iv. 5

[The Dew-berry is the *Rubus cæsius*, or heath bramble. The commentators explain this word as either raspberry or gooseberry.]

Diamond. *V.* Adamant.

DIAMOND that will receive but one form.
"Euphues' Golden Legacie."

THE third part [of Ind] toward the Septentrion is full cold; so that for pure cold and continual frost, the water becometh crystal. And upon the rocks of crystal grow the good diamonds that be of trouble colour. Yellow crystal draweth colour like oil. And they be so hard that no man may polish them. And men find many times hard Diamonds in a mass, that cometh out of gold, when men pure it and fine it out of the mine; when men break that mass in small pieces. And they grow many together, one little, another great. And they grow together male and female. And they be nourished with the dew of heaven. And they engender commonly and bring forth small children, that multiply and grow all the year. I have oft times assayed that if a man keep them with a little of the rock, and wet them with May-dew often-times, they shall grow every year; and the small will wax great. For right as the fine pearl congealeth and waxeth great of the dew of heaven, right so doth the very Diamond. And men shall bear the Diamond on his left side; for it is of greater virtue then, than on the right side. For the strength of their growing is toward the North, that is the left side of the world; and the left part of man is when he turneth his face toward the East. He that beareth the Diamond upon him, it giveth him hardiness and manhood, and it keepeth the limbs of his body whole. It giveth him victory of his enemies in play and in war (if his cause be rightful); and it keepeth him that beareth it in good wit; and it keepeth him from strife and riot, from sorrows and from enchantments, and from fantasies and illusions of wicked spirits. And if any cursed witch or enchanter would bewitch him that beareth the Diamond; all that sorrow and mischance shall turn to himself through virtue of that stone. And also no wild beast dare assail the man, that beareth it on him. Also the Diamond should be given freely without coveting and without begging; and then it is of greater virtue. And it maketh a man more strong and more sad against his enemies. And it healeth him that is lunatic, and him that the fiend pursueth or travaileth. And if venom or poison be brought in presence of the Diamond, anon it beginneth to wax moist and for to sweat.

Natheless it befalleth often time, that the good Diamond loseth his virtue, by sin and for incontinence of him that beareth it; and then it is needful to make it to recover his virtue again, or else it is of little value.

Sir John Mandeville, ch. xiv.

Dock. *V.* Burr.

KING HENRY V., v. 2, 52.

ALL kinds of Docks have this property, that what flesh or meat is sod therewith, though they be never so old, hard or tough, they will become tender and meet to be eaten.

Lupton, "A Thousand Notable Things," bk. i. § 30.

Doe. *V.* Hart, Stag, Deer.

Dog.

[Often used by Shakespeare, though it is said that he has no good word for a Dog; but *cf.* "Lear," iii. 6, 65, and "Taming of the Shrew," Induction, i. 21.]

NOTHING is more busy or wittier than an Hound, for he hath more wit than other beasts. Oft Hounds gender with wolves, and of that gendering cometh cruel Hounds. Also oft the Indians teach Bitches, and leave them in woods by night, for tigers should gender with them, and of them come most sharp Hounds and swift, and be so strong, that they throw down cruel beasts as lions. The cruelness of Hounds abateth to a meek man.

Bartholomew (*Berthelet*), bk. xviii. § 25.

Gentleness and nobility of Hounds and of Bitches is known by length of face and of the snout, and by breadth of the breast, and by smallness of the womb and flank. And a gentle Hound hath long ears and pliant, and long legs and small, and that is needful to be the more swift

in course and in running; and his tail is more long and crooked than the tails of other Hounds, and hath less flesh than a Dog and shorter hair, and more thin and smooth.

Ibid., § 26.

Under the Hound's tongue lieth a worm that maketh the Hound wood, and if this worm is taken out of the tongue then the evil ceaseth. The violence and biting of a wood Hound is so much, that his urine grieveth a man if he tread thereon, and namely if he have a botch or a wound. Also who that throweth his own urine upon the urine of a wood Hound, he shall anon feel sore ache of the guts and of the loins. Also the Hound is envious; and he gathereth herbs privily, by whom he purgeth himself with parbraking [vomiting] and casting, and hath envy, and is right sorry, if any man knoweth the virtue of those herbs. *Ibid.*, § 27.

THE tongue of a Dog laid under the great toe within the shoe doth cease the barking of Dogs at the party that so wears the same.

Lupton, "A Thousand Notable Things," bk. vii. § 22.

IF you pluck out one of the eyes of a black Dog, whiles he is living, and will carry it with you, it will make that no Dogs shall bark at you; yea, though you walk among them. But it will be more sure, if you put thereto a little of the heart of a wolf. *Ibid.*, § 85.

THE uttermost or last joint of the tail of a young whelp, after he is forty days old, being writhen off, the same Dog will never be mad. Besides that his tail will be thereby of a comely length. *Ibid.*, bk. ii. § 45.

THE teeth of a mad Dog that hath bitten a man or woman, tied in leather, and then hanged at the shoulder, doth preserve and keep the party that bears it, from being bitten of any mad Dog. *Ibid.*, bk. iv. § 52.

IF a wood Hound's drivelling fall into the water, it infecteth the water; and who that drinketh of that water shall be dropsical and wood.

Bartholomew (*Berthelet*), bk. vii. § 68.

THE delicate, neat and pretty kind of Dogs called the Spaniel-gentle, or the Comforter [Maltese Dog]:—These little Dogs are good to assuage the sickness of the stomach, being oftentimes thereunto applied as a plaster preservative, or borne in the bosom of the diseased and weak person. Moreover, the disease and sickness changeth his place and entereth (though it be not precisely marked) into the Dog, which to be truth experience can testify, for these kind of Dogs sometimes fall sick and sometimes die, without any harm outwardly enforced, which is an argument that the disease of the gentleman or gentlewoman entereth into the Dog by the operation of heat intermingled and infected.

Topsell, "Four-footed Beasts," pp. 135-6.

THAT men may appear to have Dogs' faces :—Take the fat of a Dog's ear, and anoint with it a little new silk, put it in a new lamp of green glass, and place it among the men, and they see Dogs' faces.

Albertus Magnus, "Of the Wonders of the World."

IN many places our Mastiffs (besides the use which tinkers have of them, in carrying their heavy budgets) are made to draw water in great wheels out of deep wells. Besides these also, we have Sholts [? Shoughs] or Curs daily brought out of Iceland, and much made of amongst us, because of their sauciness and quarrelling.

Holinshed, "Description of Britain," pp. 230-1.

Dog-ape. [Perhaps a **He-ape.**]

AS YOU LIKE IT, ii. 5, 27.

OF Apes some be like to an Hound in the face, and in the body like to an Ape.

Bartholomew (*Berthelet*), bk. xviii. § 96.

Dog-fish.

i. KING HENRY VI., i. 4, 107.

THE Dog-fish is a terrible monster, and hostile to all living creatures, which die from its blows. These hunt the shoals of fish in the sea, like dogs hunting wild beasts on land, except that they cannot bark ; but instead of a bark they have a horrible breath. These monsters are with difficulty killed by many fish-spears. Its gall is said to be poison, and if any one eats the quantity of a bean of it, he dies after a week. *Hortus Sanitatis*, bk. iv. ch. xvii.

Dolphin.

i. KING HENRY VI., i. 4, 107.

THE Dolphins follow man's voice, and come together in flocks to the voice of the symphony, and have liking in harmony ; and in the sea is nothing more swift than

Dolphins be. For oft they startle [spring] and overleap ships, the whose leaping and playing in the waves of the sea tokeneth tempest. And in the river of Nile is a kind of Dolphins with ridges [back-bones], toothed as a saw, that cutteth the tender wombs of crocodiles, and slayeth them. Dolphins know by the smell if a dead man that is on the sea ate ever of Dolphin's kind; and if the dead man hath eaten thereof, he eateth him anon; and if he did not, he keepeth and defendeth him from eating and biting of other fish, and shoveth him and bringeth him to the cliff with his own wroting [*cum rostris suis* (*Bartholomew*)—so wroting is rooting with the snout, as a pig does].

Bartholomew (*Berthelet*), bk. xiii. § 29.

THE Dolphin is called the brother of man, because he is in some degree like to man in his ways. They sleep on the water, so that they may be heard to snore. They live to 140 years. The Dolphin alone among fish has no gall. When a Dolphin dies, the other Dolphins come together and surround him, and bear him down to the depths, and bury him, lest other fish should eat him. Small Dolphins are always together like flocks of sheep; and they have two big Dolphins as guards. Dolphins have their eyes on their backs, and their mouths on the opposite side, and therefore they are not good at catching their prey, because of the want of agreement of the mouth and the eyes; therefore they turn their mouths towards the heaven, and their backs and eyes towards the earth, so as to follow their prey. They are said to have helped sailors when their ship was about to be wrecked. They are supposed also to weep when they are caught. *Hortus Sanitatis*, bk. iv. ch. xxvii.

Dormouse.

TWELFTH NIGHT, iii. 2, 20.

GLIRES [*i.e.*, Dormice] be little beasts, as it were great mice, and have that name (glires) for sleep makes them fat. They love their fellows that they know, and strive and fight against other. And they love their father and mother with great mildness and pity, and feed and serve them in their age. *Bartholomew* (*Berthelet*), bk. xviii. § 57.

THE soles of the feet anointed with the fat of a Dormouse doth procure sleep.

Lupton, "A Thousand Notable Things," bk. ii. § 16.

If the viper find their nest, because she cannot eat all the young ones at one time, at the first she filleth herself with one or two, and putteth out the eyes of all the residue, and afterwards bringeth them meat and nourisheth them being blind, until the time that her stomach serveth her to eat them every one. But if it happen that in the mean time, any man chance to light upon these viper-nourished-blind Dormice, and to kill and eat them, they poison themselves through the venom which the viper hath left in them. Dormice are bigger in quantity than a squirrel. It is a biting and an angry beast.

Topsell, "Four-footed Beasts," p. 409.

Dove.

He eats nothing but doves, love, and that breeds hot blood.
TROILUS AND CRESSIDA, iii. 1, 140.

WHEN the Culver [*i.e.*, Dove] hath birds [*i.e.*, young], anon the male ruleth the birds. And if the female tarry over long ere she come to the birds for soreness of the birth, then the male smiteth and beateth her, and compelleth her to sit herself upon the birds. And when the birds wax, the male goeth and sucketh salt earth; and he giveth and putteth it in the mouth of the birds, to make them have talent to meat. A Culver hath no gall, and hurteth and woundeth not with the bill, but his own peer. And hath groaning instead of song.

Bartholomew (*Berthelet*), bk. xii. § 6.

DOVES are very hot, and eat small stones to temper the stomach. The fresh flesh of a Dove helps against serpents.

Hortus Sanitatis, bk. iii. ch. xxxii.

V. Pigeon.

Dragon.

A lonely dragon, that his fen
Makes fear'd and talk'd of more than seen.
CORIOLANUS, iv. 1, 30.

THE Dragon is most greatest of all serpents, and oft he is drawn out of his den, and reseth up into the air, and the air is moved by him; and also the sea swelleth against his venom. And he hath a crest with a little mouth, and draweth breath at small pipes and strait, and reareth his tongue, and hath teeth like a saw. And hath strength, and not only in teeth, but also in his tail, and grieveth both with biting and with stinging, and hath not so much venom as other serpents; for to the end to slay any thing, to him venom is not needful; for whom he findeth he slayeth, and the elephant is not sicher of him [safe from him] for all his greatness of body, for he lurketh

in the way where the elephant goeth, and bindeth and spanneth his legs, and strangleth and slayeth him. The Dragon breedeth in Ind and in Ethiopia, there as is great burning of continual heat. The Dragon is twenty cubits great. Oft four or five of them fasten their tails together, and reareth up the heads, and sail over sea and over rivers to get good meat. The cause why the Dragon desireth his blood is coldness of the elephant's blood, by the which the Dragon desireth to cool himself. The Dragon is a full thirsty beast, insomuch that unneath [hardly] he may have water enough to quench his great thirst, and openeth his mouth therefore against the wind to quench the burning of his thirst in that wise. Therefore when he seeth ships sail in the sea in great wind, he flieth against the sail to take there cold wind, and overthroweth the ship sometimes for greatness of body, and by strong rese against the sail. And when the shipmen see the Dragon come nigh, and know his coming by the water that swelleth against him, they strike the sail anon, and scape in that wise. Also for might of the venom, his tongue is always a-reared [raised up], and sometimes he setteth the air on fire by heat of his venom; so that it seemeth, that he bloweth and casteth fire out of his mouth, and sometimes he bloweth out outrageous blasts, and thereby the air is corrupt and infected, and thereof cometh pestilent evils. And they dwell sometimes in the sea, and sometimes swim in rivers, and lurk sometimes in caves and in dens, and sleep but seld, but wake nigh always. The Dragon's biting that eateth venomous beasts is perilous, as the Dragon's biting that eateth scorpions, for against his biting unneath is any remedy or medicine found. Also all venomous beasts flee and void the grease and the fatness of the Dragon; and his grease meddled with honey cureth dimness of the eyes. Also those fishes die that be bitten of the Dragon.

Bartholomew (Berthelet), bk. xviii. § 38.

THE Dragon has wings formed from its loose and mobile skin, and they are broad in proportion to the size of its body. Wherever it stops, it poisons the air. Between eagles and Dragons there are often fights, and these much more doubtful if in the air. Also the vulture and the Dragon fight, because they prey on animals.

From the brains of Dragons is hatched the stone Dracontias; but the stone is only to be taken from the living animal; for if it die first, the hardness of the stone disappears with the breath. Dragons are put to sleep with medicated grasses, and thus the stone is procured; and the Eastern kings are especially proud of the use of this stone. The heads of Dragons make a house prosperous and fortunate. Dragon's flesh is of the colour of glass, and it cools those who eat it. Therefore the Ethiopians who dwell on that burning coast gladly eat the flesh of Dragons, so that their factors tame the Dragon with certain songs, and, sitting on his back, guide him with a bridle until they come into Ethiopia. *Hortus Sanitatis*, bk. ii. ch. xlviii.

[This last statement recalls Mr. Waterton's exploit with the alligator.]

It was wont to be said, because Dragons are the greatest serpents, that except a serpent eat a serpent, he shall never be a Dragon. In Ethiopia they grow to be thirty yards long. There are tame Dragons in Macedonia, where they are so meek, that women feed them, and suffer them to suck their breasts like little children,—their infants also play with them, riding upon them and pinching them, as they would do with dogs. The apples of their eyes are precious stones, and as bright as fire. The Africans believe that the original of Dragons took beginning from the unnatural conjunction of an eagle and a she-wolf. The Dragons of Phrygia when they are hungry turn themselves towards the west, and gaping wide, with the force of their breath do draw the birds that fly over their heads into their throats. They greatly preserve their health by eating of wild lettuce, for that they make them to vomit, and they are most specially offended by eating of apples. They renew and recover their sight again by rubbing their eyes against fennel, or else by eating of it. The Indians take a garment of scarlet, and picture upon it a charm in golden letters,—this they lay upon the mouth of the Dragon's den, for with the red colour and the gold, the eyes of the Dragon are overcome, and he falleth asleep, the Indians in the mean season watching and muttering secretly words of incantation; when they perceive he is fast asleep, suddenly they strike off his neck with an axe, and so take out the balls of his eyes, wherein are lodged those rare and precious stones which contain in them virtues unutterable. Many times it falleth out, that the Dragon draweth in the Indian both with his axe and instruments into his den and there devoureth him, in the rage whereof he so beateth the mountain that it shaketh. [*Topsell* gives several long stories of the love of some Dragons for men and women, and lastly the tale of Winckelried, who slew a horrible Dragon, whereat for joy he lifted up his sword, and the blood of the Dragon dripped off the sword and killed him.] The eagles when they shake their wings make the Dragons afraid with their rattling noise, then the Dragon hideth himself within his den. The eagle devoureth the Dragons and little serpents upon earth, and the Dragons again and serpents do the like against the eagles in the air. The griffins are likewise said to fight with the Dragons and overcome them. The panther [*q.v.*] also is an enemy unto the Dragons, and driveth them many times into their dens. *Topsell*, "History of Serpents," pp. 706-15.

The inhabitants of Paraca, by eating a Dragon's heart and liver, attain to understand the language (if so I may term it) of beasts. *Purchas*' "Pilgrims," p. 457 (ed. 1616).

Drone.

A huge feeder;
Snail slow in profit, and he sleeps by day
More than the wild-cat; drones hive not with me.
MERCHANT OF VENICE, ii. 5, 48.

The Drone is a larger kind of bee; and it eats the fruit of others' labour; for it eats what it has not worked for, as it makes no honey.

Hortus Sanitatis, bk. iii. ("Of Birds"), ch. li.

Duck.

If you see Ducks fly massed together, even though the sky be clear, you will expect rain speedily; if they flap their wings together while on the land, you may suppose that there will be a gale.

Jonston, "Natural History of Birds," ch. iii. § 3 (1657).

Eagle.

LOVE'S LABOUR'S LOST, iv. 3, 334.

Among all manner kinds of divers fowls, the Eagle is the more liberal and free of heart; for the prey that she taketh, but it be for great hunger she eateth not alone, but putteth it forth in common to fowls that follow her; but first she taketh her own portion and part. And therefore oft other fowls follow the Eagle for hope and trust to have some part of her prey. But, when the prey that is taken, is not sufficient to her self, then as a king that taketh heed of a commonty [common people], he taketh the bird that is next to him, and giveth it among the other, and serveth them therewith. And she setteth in her nest two precious stones, which be called agates; the one of them is male, and that other female; and it is said that they cannot bring forth their birds without those stones. And she layeth in her nest that precious stone, that bright agate, to keep her birds from the venomous biting of creeping worms. And among all fowls, in the Eagle the virtue of sight is most mighty and strong; for in the Eagle the spirit of sight is most temperate, and most sharp in act and deed of seeing and beholding the sun in the roundness of his circle, without any blemishing of eyes; and the sharpness of her sight is not rebounded again with clearness of light of the sun, neither disparpled [dispersed]. Also there is one manner Eagle that is full sharp of sight, and she taketh her own birds in her claws, and maketh them to look even on the sun, and that ere their wings be full grown, and except they look stiffly and steadfastly against the sun, she beateth them, and setteth them even tofore the sun; and if any eye of any of her birds watereth in looking on the sun, she slayeth him, as though he went out of kind; or else driveth him out of the nest, and despiseth him, and setteth not by him. Also the Eagle is a fowl that seldom sitteth abrood, and seldom hath birds; and nourisheth and feedeth her birds. The Eagle layeth three eggs at the most, and throweth the third egg out of the nest; for she sitteth abrood heavily thereupon. And at that time she is so much feebled, that she may not well hunt birds of other fowls; for then her claws be crooked, and her wings wax white, and then she is sore grieved in feeding of her birds. And if it happeth that the Eagle hath three birds, she throweth out one of her nest, for difficulty of feeding and nourishing; but a bird that is called ossifraga feedeth the bird that the Eagle casteth so haply out of her nest. In age the Eagle hath darkness and dimness in eyes, and heaviness in wings. And against this disadvantage she is taught by kind to seek a well of springing water, and then she flieth up into the air as far as she may, till she be full hot by heat of the air and by travail of flight, and so then by heat the pores be opened, and the feathers chafed, and she falleth suddenly into the well, and there the feathers be changed, and the dimness of her eyes is wiped away and purged, and she taketh again her might and strength. Also, when the Eagle ageth, the bill waxeth so hard and so crooked, that unneath he may take his meat. And against this disadvantage he findeth a remedy; for he seeketh a stone, against the which he smiteth and beateth strongly his bill, and cutteth off the charge of the bill, and receiveth meat and might and strength, and so becometh young again. The gentle falcon or other such fowls unneath take preys on that day that they hear the Eagle; and that perchance cometh of great dread. And that Eagle that taketh her prey on the water hath one foot close and whole, as the foot of a gander, and therewith she ruleth herself in the water, when she cometh down because of her prey. And her other foot is a cloven foot, with full sharp claws, with the which she taketh her prey. And the Eagle's feathers have a privy fretting virtue; for the Eagle's feathers done and set among feathers of wings of other birds corrumpeth and fretteth them; as strings made of wolf's guts done and put in a lute or in an harp among strings made of sheep's guts do destroy and fret and corrump the strings made of sheep's guts, if it so be that they be set among them. Also she is right cruel against her own birds, for, to teach and to compel them to take prey of other birds, she beateth and woundeth them with her bill.

Bartholomew (*Berthelet*), bk. xii. § 1.

Wherever an Eagle sees from on high a serpent, he attacks it with great clamour, and tears it with his claws, and after taking out the deadly venom from its entrails, he devours it, and the strength of the venom which was in it, being cooked by the heat of the Eagle, is extinguished. And by this experiment he is either made sad, or else he glories in it. There is in the North a large Eagle which always lays two eggs; and it catches a hare or a fox, and carefully flays off its skin, in which it wraps its eggs, and puts them in the warmth of the sun, and so leaves them and does not sit, but waits until they are broken by natural maturity, and returns when the young birds are hatched, and then feeds them until they attain to perfect strength. *Hortus Sanitatis*, bk. iii. ch. i.

HARTS that cast their horns, snakes their skins, Eagles their bills, become more fresh for any other labour.

Lilly, Prologue to "Campaspe."

THE princely Eagle, fearing to surfeit on spices, stoopeth to bite on worm-wood. *Ibid.*, Prologue to "Sapho and Phaon."

THE Eagle is never stricken with thunder.

Ibid., Act iii. Scene 3.

EAGLES cast their evil feathers in the sun.

Ibid., "Galatea," Act iii. Scene 4.

Ebony.

Black as ebony.

LOVE'S LABOUR'S LOST, iv. 3, 247.

EBONY is oft set by cradles, for black sights should not fear the children.

Bartholomew (*Berthelet*), bk. xvii. § 52.

EBONY is a tree whereof the wood is black as jet within, and beareth neither leaves nor fruit.

Batman's addition to *Bartholomew*, *l.c.*

Eel.

LOVE'S LABOUR'S LOST, i. 2, 30.

THE Eel is generated from the slime of other fishes; it is hard to skin, and very difficult to kill, as it lives even after it has been skinned; it is disturbed by the sound of thunder. It is most easily caught when the Pleiades have set. And they say that in the Eastern river Ganges, Eels are gendered with feet to walk on the land. Eels live for eight years; and they exist without water for six days while the North-east wind blows, but less while the South wind blows. Among all Eels there is no male nor female, and they gender neither live creature nor egg, as they are neuter.

Hortus Sanitatis, bk. iv. § 2.

THE Eel is a well known fish; and its virtues are wonderful; for if it dies from want of water and its body remains whole, and if strong vinegar be mixed with the blood of a vulture, and it be put somewhere under dung, they will all revive just as they were before. And if the worm [? spinal marrow] of that Eel be extracted and set in the foresaid mixture for a month, the worm will be changed into a very black Eel, and if any one eat of that Eel he will die.

Albertus Magnus, "Of the Virtues of Animals."

IN many ponds, all the water and mud taken out, by and by Eels do breed, if rain-water come into them, for that with the dew, they do live and are nourished.

Lupton, "A Thousand Notable Things," bk. iii. § 63.

IF you have many Eels in a vessel of wine, and put mulberries to them,—if any one drink of that, he will abhor wine for a year, and perhaps for ever.

Albertus Magnus, "Of the Wonders of the World."

Thunder shall not so awake the beds of eels.

PERICLES, iv. 3, 156.

Eels that never will appear
Till that tempestuous winds or thunder tear
Their slimy beds.

Marston's "Scourge of Villany," satire vii. l. 78.

V. **Fish.**

Eglantine.

MIDSUMMER NIGHT'S DREAM, ii. 1, 252.

[The sweet-briar.]

V. s. **Rose.**

Eisel. *V.* Vinegar.

HAMLET, v. 1, 299.

Elder

The stinking elder.
CYMBELINE, iv. 2, 59.

THE Ellern is a little nesh tree, and beareth flowers and fruit twice in one year, and that fruit is black with horrible smell and savour. And this is, therefore, unprofitable to eat. And wonder it is to see in Ellern, for if the middle rind of the stalk, or of the root, be shaven upward, then it purgeth upward, and if it be shaven downward, then it purgeth downward. *Bartholomew* (*Berthelet*), bk. xvii. § 144.

THE inner bark of Elder applied to any burning takes out the fire immediately. An extract or theriaca may be composed of the berries which is not only efficacious greatly to assist longevity (so famous is the story of Neander), but is a kind of catholicon against all infirmities whatever. And yet when I have said all this, I do by no means commend the scent of it, which is very noxious to the air. A certain house in Spain, seated amongst many Elder-trees, diseased and killed almost all the inhabitants, which when at last they were grubbed up became a very wholesome and healthy place. *Evelyn's* "Silva," bk. i. ch. xx. § 18.

SHEPHERDS think that pipes made from Elder are more sonorous; and it is cut when the shrub cannot hear the song of the cock. Its leaves drunk in wine resist the bites of serpents. *Hortus Sanitatis*, bk. i. ch. ccccvi.

Elephant.

The elephant hath joints, but none for courtesy.
TROILUS AND CRESSIDA, ii. 3, 113.

THESE beasts void and flee the mouse. When they be sick, they gather good herbs, and ere they use the herbs, they heave up the head, and look up toward heaven, and pray for help of God in a certain religion. If Elephants see a man coming against them, that is out of the way in wilderness, for they would not affray him, they will draw themself somewhat out of the way, and then they stint [wait], and pass little and little tofore him, and teach him the way. And if a dragon come against him, they fight with the dragon, and defend the man, and putteth them forth to defend the man strongly and mightily; and do so namely [especially] when they have young foals, for they dread that the man seeketh their foals. They dread and flee the voice of the least sound of a swine. Also between Elephants and dragons is perpetual wrath and strife [*v.* **Dragon**]. And when the Elephant sitteth, he bendeth his feet; and may not bend four at once for heaviness and weight of the body, but he leaneth to the right side, or to the left side, and sleepeth standing, and

he bendeth the hinder legs right as a man. If he hath iron in his body, oil is given him to drink, and the iron is drawn out by drink of oil. And Elephants be without gall, but they be accidentally cruel and fierce, when they be too soon angered, or if they be wine-drunken to make them sharp to fight in battle. Also no beast liveth so long as the Elephant, and his complexion is like to the air that he dwelleth in. Elephants keep lore and discipline of the stars, and in waxing of the moon go to rivers, and when they be besprung with liquor, they salute and welcome the rising of the sun with certain movings, as they may, and then they turn again into woods and launds. Their youth is known by whiteness of teeth, of the which teeth that one is always working, and that other is spared, lest he should wax dull with continual smiting and rubbing, but when they be pursued with hunters, then they smite both together, and break them, that they be no longer pursued, when the teeth be appaired and defoiled [damaged], for they know that the teeth be cause of their peril. And a cave or a ditch is made under the earth, as it were a pit-fall in the Elephant's way, and unawares he falleth therein. And then one of the hunters cometh to him, and beateth and smiteth him, and pricketh him full sore. And then another hunter cometh and smiteth the first hunter, and doth him away, and defendeth the Elephant, and giveth him barley to eat; and when he hath eaten thrice or four times, then he loveth him that defended him, and is afterwards mild and obedient to him. And if it happeth that he swalloweth a worm that is hight Chameleon, he taketh and eateth of wild olive-tree, and is so holp against the venom. Also the Elephant's bones brent [*i.e.* burnt] chase and drive away serpents and all venomous beasts. Also there is another thing that is full wonderful; for among the Ethiopians in some countries, Elephants be hunted in this wise:—There go in desert two maidens all naked and bare with open hair of the head; and one of them beareth a vessel, and the other a sword; and these maidens begin to sing alone, and the beast hath liking when he heareth their song, and cometh to them, and licketh their teats, and falleth asleep anon for liking of the song, and then the one maid sticketh him in the throat, or in the side with a sword, and the other taketh his blood in a vessel, and with that blood the people of the same country dye cloth [which is called purple—*Bartholomew*].

Bartholomew (and *Berthelet's* translation), bk. xviii. §§ 42-5.

Elephants of all other beasts do chiefly hate the mouse; so that if they shall see or perceive that a mouse hath once touched their meat that is before them, they loathe the same, and will not eat a bit thereof.

Lupton, "A Thousand Notable Things," bk. vi. § 43.

ELEPHANTS cannot bend their legs and thighs except in youth. Its inside is like a pig's inside, and therefore like a man's. It has no joints in its legs.

Hortus Sanitatis, bk. ii. ch. lv.

[Sir Thomas Browne ("Vulgar Errors," bk. iii. ch. i.) adduces sundry grave arguments to prove that an Elephant has joints:]

WHILE men conceive they never lie down, and enjoy not the position of rest ordained unto all pedestrious animals, hereby they imagine (what reason cannot conceive) that an animal of the vastest dimension and longest duration, should live in a continual motion, without that alternity and vicissitude of rest whereby all others continue.

"Vulgar Errors," bk. iii. ch. i.

IN the woods or fields where they [Indians or Africans] suspect [Elephants'] teeth to be buried, they bring forth pots or bottles of water, and disperse them, here one, there another, and so let them stand, and tarry to watch them,—so one sleepeth, another singeth, or bestoweth his time as he pleaseth; after a little time they go and look in their pots, and if the teeth lie near their bottles, by an unspeakable and secret attractive power in nature, they draw all the water out of them that are near them, which the watchman taketh for a sure sign, and so diggeth about his bottle, till he find the tooth. [*Topsell* decides after argument that tusks are not horns.] The trunk hath two passages,—one into the head and body by which he breatheth, and the other into his mouth. It is false that they have no joints or articles in their legs. They drink not wine, except in war, when they are to fight, but water at all times, whereof they will not taste, except it be muddy and not clear, for they avoid clear water, loathing to see their own shadow therein. In the summer-time they choose out and gather the sweetest flowers, and being led into their stables, they will not eat meat until they take of their flowers, and dress the brims of their mangers therewith, pleasing themselves with their meat, because of the savour of the flowers stuck about their cratch, like dainty-fed persons which set their dishes with green herbs, and put them into their cups of wine. They are never so fierce, violent, or wild, but the sight of a ram tameth and dismayeth them, for they fear his horns; and not only a ram, but also the gruntling clamour or cry of hogs. Lions set upon the young calves of Elephants and wound them, but, at the sight of the mothers, the lions run away, and when the mothers find their young ones embrued in their own blood, they themselves are so enraged, that they kill them, and so retire from them, after which time the lions return and eat their flesh. In the River Ganges there are blue worms of sixty cubits long having two arms; these, when the Elephants come to drink in that river, take their trunks in their hands and pull them off. At the sight of a beautiful woman [Elephants] leave off all rage and grow meek and gentle. In Africa there are certain springs of water, which, if at any time they dry up, by the teeth of Elephants, they are opened and recovered again. In the night-time, Elephants seem to lament with sighs and tears their captivity and bondage, but if any come to that speed, like modest persons they refrain suddenly, and are ashamed to be found either murmuring or sorrowing. When they drink a leech, they are grievously pained. The fime [or dung] by anointing cureth a lousy skin, and taketh away that power which breedeth these vermin; the same perfumed driveth gnats or marsh-flies out of a house.

Topsell, "Four-footed Beasts," pp. 150-65.

Elm.

The barky fingers of the elm.
MIDSUMMER NIGHT'S DREAM, iv. 1, 49.

THE shadow of Elms is mild and nourishing to those things that it falls upon. *Hortus Sanitatis*, bk. i. ch. dvii.

[Evelyn ("Silva," bk. i. ch. iv.), among the uses of the Elm, states that it is proper for dressers and shovel-board tables, and that cattle prefer the dried leaves to oats in the winter when hay and fodder are dear.]

THE leaves of an Elm-tree, or of peach-tree, falling before their time doth foreshow or betokens a murrain or death of cattle. *Lupton*, "A Thousand Notable Things," bk. iii. § 25.

Emerald.

MERRY WIVES OF WINDSOR, v. 5, 74. [As adjective.]

OF all green precious stones is the chief. Men in old time gave thereto the third dignity after margarites and unions. In no herbs nor in precious stone is more greenness than in the stone Emerald. It passeth herbs and grass, twigs and branches. And it infecteth the air about it with passing green colour. And his green colour abateth not in the sun in no manner wise. Nothing comforteth more their eyes that be gravers than this stone. Thereof be twelve manner of kinds, but the most noble be found in Scythia, and in Bactria holdeth the second place. And Emeralds be found among and under stones, and in chines thereof, when the Northern wind bloweth. For then the earth is uncovered, and Emerald shineth among the stones. For in such wind gravel and sand is most moved. Though the Emerald be green by kind, yet if it be meddled with wine or with oil, his green colour increaseth. This stone is taken of and from griffins, and plenty of Emeralds may not be found, for great griffins let the coming of men by the way that goeth thereto. The body thereof hath of gift of kind a goodness of virtue to heal divers sicknesses and evils. It increaseth riches, and maketh men have good words and fair evidence in cause and in plea. If this stone be hanged about the neck, it maketh good mind, and helpeth also against all phantasies and japes of fiends, and ceaseth tempest. It is said that it helpeth them that use to divine and guess what shall befall.

Bartholomew (*Berthelet*), bk. xvi. § 88.

A TRUTHFUL and curious experimenter coming from Greece said that this stone is generated in the rocks which are under the sea, and is there found. And this is reasonable, for it is generated in veins of brass, and that evidently, but it does not come to the substance of brass; because it has the greenness of the rust of brass [*i.e.*, verdigris]. The Emerald if put in drink is suitable for deadly venoms, and for venomous bites and punctures of stings. The Emerald, if it be worn, increases substance; causes persuasiveness in all business; makes men chaste and cheerful of body and of speech; and helps in tempests. Also it makes the memory good.

Hortus Sanitatis, bk. v. § 113.

If any one carries an Emerald under his tongue, straightway he will prophesy.

Albertus Magnus, "Of the Virtues of Stones."

Eringo or Eryngo (*i e.*, Sea-holly).

Merry Wives of Windsor, v. 5, 23.

The roots condited or preserved with sugar, as hereafter followeth, are exceeding good to be given unto old or aged people that are consumed and withered with age, and which want natural moisture; they are also good for other sorts of people that have no delight or appetite to venery, nourishing and restoring the aged, and amending the defects of nature in the younger.

The manner to condite Eringoes:

Refine sugar fit for the purpose, and take a pound of it, the white of an egg, and a pint of clear water; boil them together and skim it, then let it boil until it be come to good strong syrup, and when it is boiled, as it cooleth, add thereto a saucer-ful of Rose-water, a spoon-ful of cinnamon-water, and a grain of musk, which have been infused together the night before, and now strained; into which syrup being more than half cold, put in your roots to soak and infuse until the next day;—your roots being ordered in manner hereafter following:—These your roots being washed and picked, must be boiled in fair water by the space of four hours, until they be soft, then must they be pilled clean, as ye pill parsnips, and the pith must be drawn out at the end of the root; and if there be any whose pith cannot be drawn out at the end, then you must slit them, and so take out the pith; these you must also keep from much handling, that they may be clean; let them remain in the syrup till the next day, and then set them on the fire in a fair broad pan until they be very hot, but let them not boil at all; let them there remain over the fire an hour or more, removing them easily in the pan from one place to another with a wooden slice. This done, have in a readiness great cap or royal papers, whereupon you must straw some sugars, upon which lay your roots after that you have taken them out of the pan. These papers you must put into a stove or hot house to harden; but if you have not such a place, lay them before a good fire. In this manner if you condite your roots, there is not any that can prescribe you a better way. And thus you may condite any other root whatsoever, which will not only be exceeding delicate, but very wholesome. They report that the herb Sea-holly, if one goat take it into her mouth, it causeth her first to stand still, and afterwards the whole flock, until such time as the shepherd take it forth of her mouth. *Gerard's* "Herbal," *s.v.*

[References to the rejuvenating power of Eringoes, especially when candied (or condite), occur in very many old plays.]

Estridge.

i. KING HENRY IV., iv. 1, 98.
ANTONY AND CLEOPATRA, iii. 13, 197.

[Estridge, as Douce suggests, is probably a goshawk or Estridge-falcon; but the word was also commonly used for ostrich.]

Ewe. *V.* Sheep.

Falcon.

[Gervase Markham, in his treatise on Husbandry, devotes twenty-six chapters to the treatment of the diseases of Hawks. In the last chapter he says: "It is a known experience among the best falconers, that if the Gerfalcon shall lose but two or three drops of blood, it is mortal, and the Hawk will die suddenly after; which to prevent, if the blood proceed from any pounce, which is most ordinary, then upon the instant hurt, you shall take a little hard merchant's wax, and drop it upon the sore, and it will presently stop it."]

THE Falcon is a royal fowl, and desireth prey, and useth to sit on his hand that beareth him, and is a bold bird and an hardy, as is the goshawk. And hath little flesh in comparison to his body, and hath many feathers; and therefore he is more light to fly. And is so great-hearted that if he fail of his prey in the first flight and rese, in the second he taketh wreak on himself; and so if he be wild, unneath that day he seeketh prey; and if he be tame, as it were for shame he flieth about in the air, and then unneath he cometh to his lord's hands. For he holdeth himself overcome, and as it were put out of kind, if he taketh not the fowl that he flieth to. And among all birds and fowls, these fowls have little affection, and take little heed of their birds [*i.e.*, young]. With the same office of business that he feedeth his own birds, with such service he taketh and feedeth the birds that the eagle throweth out of her nest, and is unknown to him. *Bartholomew* (*Berthelet*), bk. xii. § 20.

V. **Hawk—Goshawk.**

Fawn.

As You Like It, ii. 7, 128.

THE hart-calf hight *hinnulus*, and is the hart's son, and hath that name *hinnulus* of becking or nodding, for he is hid by becks and signs of his mother, and is a feeble beast and loath to fight, and he is most sharp of sight, and swift of course and of running. And the hart-calf is contrary to the serpents in a wonder wise; for he that is anointed with his suet or with his blood shall not be touched of any serpent. *Bartholomew* (*Berthelet*), bk. xviii. § 59.

Fennel.

Hamlet, iv. 5, 180.

FENNEL is a common herb, and is of great virtue and might, and is hot and dry in the second degree, and hath virtue to temper and to shed, and to open, and to carve, and to cut; and that by subtle cause and qualities thereof. The juice of the stalk and of the root thereof sharpeth the sight. And it is said that serpents taste thereof and do away the age of their years. Serpents make this herb noble, and they restore the sight, and maketh it sharp with juice thereof. And understanding of inwit is arred [*i.e.*, increased] therewith, and dimness put off. The seed thereof drunk with wine helpeth against biting of serpents, and stinging of scorpions. The root thereof, if it be sod in wine, healeth biting of hounds. *Bartholomew* (*Berthelet*), bk. xvii. § 70.

Ferret.

JULIUS CÆSAR, i. 2, 186.

THERE is no beast that more desireth fish than Ferrets and cats, and yet I cannot consent unto them which will have the Ferret descend and hunt fish in the waters like otters or beavers. Young boys and scholars also use to put them into the holes of rocks and walls to hunt out birds, and likewise into hollow trees, whereout they bring the birds in the claws of their feet. Wheras a long fly (called a Friar) flying to the flaming candles in the night, is accounted among poisons, the antidote and resister thereof is a goat's gall or liver mixed with a Ferret, or wild weasel, and the gall of Ferrets is held precious against the poison of asps, although the flesh and teeth of a Ferret be accounted poison. *Topsell*, "Four-footed Beasts," pp. 170-1.

Fig.

MIDSUMMER NIGHT'S DREAM, iii. 1, 170.

The fig of Spain.
KING HENRY V., iii. 6, 62.

THE Fig-tree is more fruitful than other trees, for it beareth fruit three or four times in one year, and while one ripeth, another springeth anon. And the stock thereof done in water sinketh anon to the ground, and riseth and cometh up above the water after that it hath lien in the mud, against the common course of kind. Tofore Pythagoras' time hawks were fed with Figs, tofore he brought them to choose of flesh, that is the stronger meat. Figs do away rivels [*i.e.*, wrinkles] of old men, if they eat thereof among their meat. And full cruel bulls become mild anon if they be tied to a Fig-tree. The milk of the Fig-tree hath virtue of running together to make cheese. Some trees shall be set nigh trees that bear well fruit, that blasts of wind may be borne therefrom to the Fig-tree, and thereto the southern wind is better than the northern wind, for the northern wind grieveth the Fig-tree more than the southern wind. *Bartholomew* (*Berthelet*), bk. xvii. § 61.

THE blood made from Figs is not good, and for this reason it makes lice. *Hortus Sanitatis*, bk. i. ch. cxciv.

I DO look now for a Spanish Fig, or an Italian salad daily.

Webster, "Vittoria Corrombona."

Poisoned! A Spanish Fig
For the imputation.

Ibid., "Duchess of Malfi," ii. 3.

A LADDER made of the wood of a Fig-tree hath a marvellous property; for if flesh in the seething thereof be often stirred therewith, or if it be in the pot while it is seething, it makes the flesh to be sooner sodden.

Lupton, "A Thousand Notable Things," bk. x. § 98.

Filbert.

TEMPEST, ii. 2, 175.

A CATALOGUE of the best Filberts:

White } Avelans.
Red }
Large Hazel.
Long, Thin, and Great Round Nuts.

Evelyn, "Kalendarium Hortense."

THEY engender much ventosity, if they be ate with the small skins; therefore to take away the grief, it is good to blanch them in hot water. The skin thereof meddled with honey helpeth against falling of hair, and maketh hair grow in the body. *Bartholomew* (*Berthelet*), bk. xvii. § 109.

A SUPERSTITIOUS notion prevails with the common people, that if it rains about the time of Midsummer Eve, the Filberts will be spoiled that season.

Brand's "Popular Antiquities," vol. i. p. 253.

IF it be rubbed on the heads of boys who have eyes of different colours, it takes away the diversity. It helps against venom and bites, and especially with figs and rue against the punctures of scorpions.

Hortus Sanitatis, bk. i. ch. cccxiii.

Fire-drake [Will-o'-th'-Wisp].

KING HENRY VIII., v. 4, 45.

THAT which is spoken of the poison of dragons infecting the air wherein they live is to be understood of the meteor called a Fire-drake, which doth many times destroy the fruits of the earth, seeming to be a certain burning fire in the air, sometime on the sea, and sometime on the land.

Topsell, "History of Serpents," p. 713.

How many oaths flew toward heaven,
Which ne'er came half-way thither, but, like Fire-drakes,
Mounted a little, gave a crack, and fell.

Middleton, "Your Five Gallants," iii. 2.

IT may be, 'tis but a glow-worm now,—but 'twill
Grow to a Fire-drake presently.

Beaumont and Fletcher, "Beggar's Bush," v. 1.

Fish.

FISH licketh the earth and watery herbs, and so get they meat and nourishing. Also they be called *Reptilia*, creeping, because in swimming they seem as they did creep; for in swimming they creep, though they sink down to the bottom. Also Fish love their children, and feed and nourish them long time. All Fish feed and keep their birds [*i.e.*, young ones], out take frogs. Some Fish be gendered, without eggs or peasen [spawn], of slime and of ooze, of gravel and of rottenness that is upon the water. And there is a Fish that hight a Lamprey, that of his like conceiveth not, but of an adder, which he calleth to love with hissing. And therefore fishers call it with hissing and whistling, and taketh her in that wise. Unneath she dieth, though she be smit with a staff; and if she be smit with a rod, she dieth anon. It is certain that the soul of the Fish is in the tail; for unneath she is slain, though she be smit on the head; and if she be smit on the tail, or if the tail be smit off, she dieth anon. And the contrary is of the serpent, for if the head be broke and bruised or cut off, the serpent dieth anon; and if the tail be smitten, he liveth long time. Also the serpent doth away his venom ere he gender with the lamprey. Also Fish conceive of dew only as oysters and other shell-fish. Fish that be called Elich come out of the water by night, and conceive in land of the morrow dew, and bring forth their brood ; and in waning of the moon their shells be void. Also Fish is stirred to conceive and to breed by rising and down-going of stars. So Fish that hight Australis arise, when the stars that hight Pleiades begin to go down, and be not seen till Pleiades arise again. Generally Fish be gluttons, and covet much meat. And generally Fish travail more by day than by night, and more tofore midnight than after. And therefore they be hunted tofore the sun rise, and then fishers set their nets ; for that time fish see not. Full well they see when light increaseth ; but by night they seek their meat by smelling. Also there is some kind of great huge Fish with great bodies and huge, as it were mountains and hills; such was the whale that swallowed Jonas the prophet ; his womb was so great that it might be called hell. The barnacle [*q.v.*] when he knoweth and feeleth that tempest of wind and weathers be great, he cometh and taketh a great stone, and holdeth him fast thereby, as it were by an anchor, lest he be smitten away, and thrown about with waves of the sea. And so he saveth not himself by his own strength, but helpeth to save himself by heaviness and weight that is not his own. And is made steadfast and stable against the coming of tempest and storm. And shipmen see this and beware that they be not overset unwarily with tempest and with storms. Heads of salt Fish burnt healeth the biting of a wood hound, and the stinging of a scorpion. Also the juice of every Fish helpeth against venom that is drunken, and against venomous stingings. [*Cooper* ("Thesaurus," *s.v.*) adds "The mugil [*i.e.* mullet] is of all scaled Fishes the swiftest, of colour white, having a great belly, and in greediness unsatiable ; when he is full he lieth still in one place, and being afraid hideth his head, deeming thereby that no part of his body is seen. They are so desirous each of other's kind, that when fishers hang a male of that sort on their line, all the females resort unto it, and so be taken ; and likewise do all the males to the females."] Also of a Fish which hight *estaurus ;* for among Fish only that Fish cheweth his cud:

and this Fish is right witty. For when he knoweth that he is entered, and is within the dangers of the fishers' gin, he reseth not forth headlong, neither putteth his head between the rolls of the gin; but he beateth fast on the other side with his tail, and beginneth to make him a way with breaking and renting of rods, and so passeth backward. And if it happen that another Fish of the same kind seeth this doing, and how he travaileth for to break out, he busieth to help him, and taketh his tail in his mouth, and helpeth as he may to draw him out, and deliver him of the gin.

Bartholomew (*Berthelet*), bk. xiii. § 26.

V. also **Crab, Oyster, Lamprey, Whale, Conger,** etc.

BEGGARS are sometimes carted in pairs of paniers or in dossers like fresh Fish from Rye, that comes on horseback.

Dekker's "Lanthorn and Candlelight," ch. viii.

HEAVEN is not pleased with our vocation. I speak it to my grief, and to the burthen of my conscience, we fry our Fish with salt butter.

Marston's "Dutch Courtezan," ii. 3, 12.

HE, excellent in love as the sea-inhabitant,
Of whom 'tis writ that, when the flattering hook
Has struck his female, he will help her off,
Although he desperately put on himself,
But if he fail, and see her leave his eye,
He swims to land, will languish, and there die—
Such is his love to me.

Robt. Davenport, "City Nightcap," Act v. (1624).

IN Snowdonie are two lakes, whereof one beareth a movable island, which is carried to and fro as the wind bloweth; the other hath three kinds of Fishes in it, as eels, trouts and perches; but herein resteth the wonder, that all those have but one eye apiece only, and the same situate in the right side of their heads.

Holinshed, "Description of Britain," p. 129 (ed. 1586).

THE osprey, both alive and dead, yea even her very oil is a deadly terror to such Fish as come within the wind of it.

Ibid., p. 227.

THE pike is friend unto the tench as to his leech and surgeon; for when the fishmonger hath opened his side and laid out his rivet and fat unto the buyer, for the better utterance of his ware, and cannot make him away at that present, he layeth the same again into the proper place, and sewing up the wound, he restoreth him to the pond where tenches are, who never cease to suck and lick his grieved place, till they have restored him to health, and made him ready to come again to the stall, when his turn shall come about. It is believed with no less assurance of some than that an horse-hair, laid in a pail full of the like [*i.e.*, fenny] water, will in short time stir and become a living creature.

Holinshed, "Description of Britain," pp. 223-4.

THE Lomond Lake [hath] fleeting isles and Fish without fins. *Ibid.*, p. 88.

[IN Cuba] fishermen after a strange fashion used to hunt Fish, and take them by the help of another Fish, which they kept tied in a cord by the boat's side, and when they espied a Fish loosed the cord; this hunting Fish presently lays hold on the prey, and, with a skin like a purse growing behind her head, graspeth it so fast that by no means it can be taken from her, till they draw her up above the water. *Purchas*' "Pilgrims," p. 904 (ed. 1616).

Fitchew.

A dog, a mule, a cat, a fitchew, a toad, a lizard, an owl, a puttock, or a herring without a row.

TROILUS AND CRESSIDA, v. 1, 67.

V. Pole-cat.

THE skin is stiff, harsh and rugged in handling, and therefore long lasting in garments, yet the savour of it is so rank, that it is not in any great request, and moreover it offendeth the head, and produceth ache therein, and therefore it is sold cheaper than a fox-skin.

Topsell, "Four-footed Beasts," p. 172.

Flax.

MERRY WIVES OF WINDSOR, v. 5, 159.

THE Flax is made to knots and little bundles, and so laid in water, and lieth there long time. And then it is taken out of the water, and laid abroad till it be dried, and turned and wended [*Bartholomew—desiccantur*] in the sun, and then bound in pretty niches and bundles; and afterward knocked, beaten and brayed, and carfled [? sliced ?] rodded, and gnodded [? from gnide, *i.e.*, rub ?] ribbed and heckled, and at the last spun. Then the thread is sod and bleached, and bucked and oft laid to drying, wetted and washed, and sprinkled with water until that it be white, after diverse working and travail.

Bartholomew (*Berthelet*), bk. xvii. § 97.

FLAX ought by law to be sown in every country-town in England, more or less; but I see no success of that good and wholesome law.

Holinshed, "Description of England," p. 111.

Flea.

i. KING HENRY IV., ii. 1, 16, 23.

THE Flea is a little worm, and grieveth men most, and it is namely fed with powder [*i.e.*, dust]; and is a little worm of wonder lightness, and scapeth and voideth peril with leaping, and not with running, and waxeth slow, and faileth in cold time, and in summer time it waxeth quiver and swift. And the Flea is bred white, and changeth as it were suddenly into black colour, and desireth blood, and doth let them that would sleep with sharp biting, and spareth not kings, but a little Flea grieveth them, if he touch their flesh. And to Fleas wormwood is venom, and so be leaves of the wild fig-tree. And coloquintida a weed that is like to a wild nep helpeth against Fleas, if it be stamped and meddled with water, and sprung in the place there as many Fleas be; and they bite full sore against rain.

Bartholomew (*Berthelet*), bk. xviii. § 89.

A SLUTTISH kept house breedeth Fleas, and lodging next to stables of horses; also the horse-urine breedeth Fleas, his dung, falling upon his tail, breedeth snakes, his flesh, wasps.

Batman's addition to *Bartholomew*, bk. xviii. § 89.

A Flea divided in two parts revives. If the water in which brambles have been boiled be sprinkled in a house, Fleas will be quite destroyed. When the blood of a goat is placed in a hole in a house, Fleas collect about it, and then die. *Hortus Sanitatis*, bk. ii. § 118.

When any draws nigh toward their death, and their members lack blood and vital heat: then Fleas and lice leave them quite, or else draw to that part of the body where the said heat tarries the longest; which is in the hole in the neck under the chin, etc. This is a token that death is at hand.

Lupton, "A Thousand Notable Things," bk. iii. § 75.

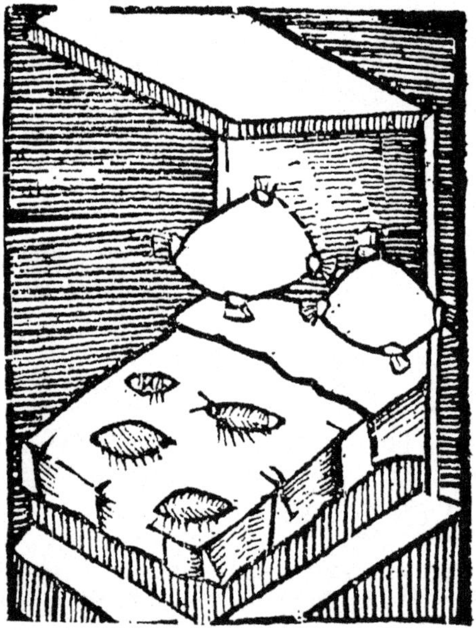

Item, paid to goodwife Wells for salt to destroy the Fleas in churchwardens' pew, 6d.

Parish Accounts of St. Margaret's, Westminster (1610).

It is not any disgrace to a man to be troubled with Fleas, as it is to be lousy. Their first original is from dust. *Mouffet*, "Theatre of Insects," p. 1102.

If any one be anointed with the milk of an ass, all the Fleas in the house will gather together upon him.

Albertus Magnus, "Of the Wonders of the World."

Fly and Flesh-fly.

When the splitting wind
Makes flexible the knees of knotted oaks,
And flies fled under shade.

Troilus and Cressida, i. 3, 51.

Flies are unquiet, and importunate, and malicious, stinging and worrying. Flies, like bees, if killed in water, sometimes revive after an hour. If Flies be burnt, and smeared with honey on bald places, they produce hair.

Hortus Sanitatis, bk. iii. ("De Avibus") § 81.

When thou wilt drive away Flies from any place that there shall none be seen there again, make the image of a Fly in the stone of a ring; or in a plate of brass or copper, or of tin, make the image of a Fly, of a spider, or of a serpent, the second face of Pisces then ascending. And whiles you are making a graving of them, say: This is the image which doth clean rid all Flies for ever. Then bury the same in the midst of the house, or hang it in any place of the house, (but if thou hast four such plates, and bury them or hang them in four corners of the house, or hide them within the walls, that nobody take them away, it were far better). But this laying of them must be when the first face of Taurus doth ascend. And so no Fly will come in there, nor tarry there. Ptolomy saith that he saw the trial hereof in the house of King Adebarus; who was very wise, and was marvellous expert in natural magic, in whose palace or place, there was neither Fly nor any other hurting worm. And that I might search it out (saith he), I brought in thither live Flies, which presently died.

Lupton, "A Thousand Notable Things," bk. ii. § 21.

In the common place where the Censors of Venice sit, there never enter any Flies. And in the flesh shambles of Toledo in Spain, is not seen but one Fly in all the whole year. And in Westminster Hall, in the timber-work, there is not to be found one spider, nor a spider's web. Because (as it is thought) the timber wherewith the roof is builded was brought out of Ireland, and did grow there. [N.B. This belief was still held in Wales in 1869.] In all which country of Ireland, I have not only heard it credibly told that there is neither spider, toad, nor any other venomous thing; but also that some of the earth of that country hath been brought hither, whereon a toad being laid, she hath died presently. Though this be marvellous and strange, yet it is true.

Lupton, "A Thousand Notable Things," bk. iv. § 31.

If you rub slightly any kind of beast or cattle with the juice of gourds in hot weather, no kind of Flies will then hurt or molest them, nor yet annoy them. A thing desired of many, and very necessary for such as rides in the hot weather.

Ibid., bk. v. § 42.

Flies are generated two ways,—by coupling with their own species, or by the putrefaction of other things. When the Flies bite harder than ordinary, making at the face and eyes of men, they foretell rain or wet weather. Trouts are taken with the Ground-fly, but chiefly with the Dung-fly; so that the anglers use to fasten one or two of them to their hook, and with a sporting or rather cunning snatching back of their line do invite the trouts more greedily to bite, and the bait being swallowed down to hang the surer. Others put as many of those Flies upon their hooks as they will hold, and plunge them quite down to the bottom, especially where they know the greater trouts use to haunt. But every month must have his several Fly; the which the fishers do very well know, who in defect of the natural Fly do substitute artificial Flies made of wool, feathers, or divers kind of silken colours, with which they cozen and deceive the fish. Only you must take heed that as soon as ever they bite, you pull your line to you, lest the fish refusing the unsavoury bait get away. We conclude this art of making Flies to be very ancient, and derived to us by long tract of time; however, we have some bold bragging book-men at this day [*i.e.*, 1582] that ascribe it to their own invention.

Mouffet, "Theatre of Insects," pp. 932, 944, 946.

As to Flies, we have none that can do hurt or hindrance naturally unto any. The cut- or girt-waisted (for so I English the word *Insecta*) are the hornets, wasps, bees, and such like, whereof we have great store, and of which an opinion is conceived that the first do breed of the corruption of dead horses, the second of pears and apples, and the last of kine and oxen ; which may be true, especially the first and latter in some parts of the beast, and not their whole substances, as also in the second, sith [since] we have never wasps, but when our fruit beginneth to wax ripe. Yet sure I am of this that no one living creature corrupteth without the production of another ; as we may see by ourselves, whose flesh doth alter into lice ; and also in sheep for excessive numbers of flesh-flies, if they be suffered to be unburied.

Holinshed, "Description of England," p. 228.

Fowl.

OUT of the fig-tree there comes such a sharp vapour, that if a hen be hanged thereon, it will so prepare her, that she will be soon and easily roasted. And the like will be if the feathers be plucked off from Fowls and birds, and the skins pulled off from beasts, and then laid or covered a day or two in a heap of wheat.

Lupton, "A Thousand Notable Things," bk. iv. § 19.

YOU are now in Lincolnshire, where you can want no Fowl, if you can devise means to catch them.

Lilly, "Galatea," Act i. Scene 4.

V. Bird.

Fox.

A Fox hight *vulpes*, and hath that name as it were wallowing feet aside [uneven-legged : see below], and goeth never forthright, but alway aslant, and with fraud. And is a false beast and deceivable ; for when him lacketh meat, he feigneth himself dead, and then fowls come to him, as it were to a carrion, and anon he catcheth one and devoureth him. The Fox halteth alway; for the right legs be shorter than the left legs; his skin is right hairy, rough, and hot; his tail is great and rough; and when an hound weeneth to take him by the tail, he taketh his mouth full of hair, and stoppeth it. The Fox doth fight with the brock for dens, and defileth the brock's den with his urine and with his dirt, and hath so the mastery over him with fraud and deceit, and not by strength. The hart is friend to a Fox, and fighteth therefor with the brock, and helpeth the Fox. The Fox is a stinking beast and corrupt, and doth corrupt oft the place that they dwell in continually, and maketh them to be barren. His biting is somedeal venomous. And when hounds do pursue him, he draweth in his tail between his legs, and when he seeth he may not scape, he [micturates] in his tail that is full hairy and rough, and swappeth his tail full of [urine] in the hounds' faces that pursue him. And the stench of the [urine] is full grievous to the hounds, and therefore the hounds spare him somewhat. The Fox feigneth himself tame in time of need; but by night he waiteth his time, and doeth shrewd deeds. And although he be right guileful in himself and malicious; yet he is good and profitable in use of medicine. For if a man have upon him a Fox-tongue in a ring or in a bracelet, he shall not be blind, as witches mean.

Bartholomew (*Berthelet*), bk. xviii. § 114.

You may take Foxes with this oil following: Anoint the soles of your shoes, with a piece of fat swine's flesh as broad as your hand, newly toasted or a little broiled at the fire, when you go out of the wood homeward. And in every of your steps, cast a piece of the liver of a swine roasted, and dipped in honey, and draw after your back the dead carcase of a cat, and when the Fox following thee comes near unto the steps, be sure to have a man nigh thee with bow and shafts to shoot at him: or by some other means to hit him. Mizaldus had this of an expert hunter.

Lupton, "A Thousand Notable Things," bk. vi. § 21.

Foxes being sod or cut in pieces, and then given to hens or geese among their meat, it makes them safe from being hurt of any Foxes after, for the space of two months (*Mizaldus*).

Ibid., bk. vii. § 44.

THE Fox takes the juice which flows from the pine-tree into his food, and so recovers his health and prolongs his life. When hungry, he imitates the barking of a dog.

Hortus Sanitatis, bk. ii. ch. clix.

SERPENTS, apes and Foxes, and all other dangerous, harmful beasts have small eyes, but sheep and oxen, which are simple, very great eyes. The Fox with his breath draweth field-mice out of their holes, like as a hart draweth out serpents with his breath, and devoureth them. In Arabia and Palestine they are so ravenous that in the night they fear not to carry into their dens old shoes and vessels, or instruments of husbandry. But if a Fox eat any meat wherein are bitter almonds [or aloes] they die thereof if they drink not presently. If wild rue be secretly hung under a hen's wing, no Fox will meddle with her. In some places they take upon them to take him [the Fox] with nets, which seldom proveth, because with his teeth he teareth them in pieces. The *French* have a kind of gin to take by the legs, and I have heard of some which have found the Fox's leg in the same gin, bitten off with his own teeth from his body; other have counterfeited themselves dead, restraining their breath and winking, not stirring any member when they saw the hunter come to take them out of the gin [and] so soon as the Fox perceiveth himself free, away he went, and never gave thanks for his deliverance. With his tail he draweth fishes to the brim of the river, and when that he observeth a good booty, he casteth the fishes clean out of the water upon the dry land, and then devoureth them. The tongue [of a Fox] either dried or green, laid to the flesh wherein is any dart or other sharp head, it draweth them forth violently. The liver dried and drunk cureth often-sighing.

Topsell, "Four-footed Beasts," pp. 174-9.

A Fox will not touch any cocks, hens, or such like pullen, that have eaten (before) the dried liver of a Reynard, nor those hens which a cock, having a collar about his neck of a Fox-skin, hath trodden.

Holland's Pliny, bk. xxviii. ch. xx.

V. also **Brock.**

Frog.

Eye of newt and toe of frog,
Wool of bat and tongue of dog.
MACBETH, iv. 1, 14.

[Note that the Frog occurs here among other animals supposed to be venomous as an ingredient of the witches' cauldron.]

THE Frog crieth greedily and maketh much noise in those marais [*i.e.*, marshes] where he is bred. And some Frogs be water-frogs; and some be of moors and of marais. And there is a manner Frog, that maketh an hound still and dumb, if he cometh in his mouth. And the Frog hath his own voice, and maketh not that voice but only in water. His eyes shineth as a candle, and namely by night. And all fish nourisheth and feedeth his brood, out-take[n] the Frog. Then the Frog is watery and moorish, crying and slimy, with a great womb and speckled there-under, and is venomous, and abominable therefore to men and most hated, and both in water and in land he liveth.

Bartholomew (*Berthelet*), bk. xviii. § 91.

BY Frogs I understand not such as arising from putrefaction are bred without copulation, and because they subsist not long are called *Temporaria.*

Sir Thos. Browne, "Vulgar Errors," bk. iii. ch. xiii.

FORBEAR in plenty of other meat this wanton eating of Frogs, as things perilous to life and health. They which use to eat Frogs fall to have a colour like lead. They did burn the young Frogs, putting the powder thereof into a cat, whose bowels were taken out, then roasting the cat, and after she was roasted, they anointed her all over with honey, then laid her by a wood-side; by the odour and savour whereof, all the wolves and foxes lodging in the said wood were allured to come to it, and then the hunters lying in wait did take, destroy and kill them. The flesh of Frogs is good against the biting of the sea-hare, the scorpion, and all kind of serpents. The broth taken into the body with roots of sea-holm expelleth the salamander. The little Frogs are an antidote against the Toads and great Frogs. *Topsell*, "History of Serpents," pp. 722-3.

V. **Paddock, Toad.**

THAT a woman may confess what she has done :—Catch a live Water-frog, and take out its tongue, and put the Frog back into the water, and put the tongue over the region of the woman's heart while she is asleep, and when she is questioned, she will tell the truth.

Albertus Magnus, "Of the Wonders of the World."

FROGS abound where snakes do keep their residence.

Holinshed, "Description of England," p. 228.

Fumitory.

Rank fumitory

KING HENRY V., v. 2, 45.

FUMITORY [*fumus terræ*] springeth and groweth out of the earth in great quantity, as smoke doth, or fumosity that cometh of the earth. And the more green the herb is, the better it is; and is of no virtue when it is dry. And is an herb with horrible savour and heavy smell, and is natheless most of virtue. For it cleanseth and purgeth melancholia, phlegm and cholera.

Bartholomew (*Berthelet*), bk. xvii. § 69.

DOVES are delighted with it.

Minsheu's Dictionary, *s.v.*

Furze.

TEMPEST, iv. 1, 180.

Is full bitter to man's taste, and is a shrub that groweth in a place that is forsaken, stony and untilled.

Bartholomew (*Berthelet*), bk. xvii. § 80.

CAMMOCK [or rest-harrow or Ground-furze] hath this singular virtue, that it gendereth fire of itself, for when the leaves thereof fall and be dry, those leaves by a little blast of hot wind and drought are set on fire.

Ibid., bk. xvii. § 138.

Gall.

Let there be gall enough in thy ink.
TWELFTH NIGHT, iii. 2, 52.

THERE breedeth on the leaves [of the oak] a manner thing sour and unsavoury. And physicians call it Gall.

Bartholomew (*Berthelet*), bk. xviii. § 134.

IF the inner part of the Gall be taken and put on a decayed tooth, it allays the pain of it.

Hortus Sanitatis, bk. i. § 206.

Garlic.

MIDSUMMER NIGHT'S DREAM, iv. 2, 43.
WINTER'S TALE, iv. 4, 162.

MEN that must needs pass by stinking places, or make clean uncleanly rotten places, arm and defend themselves with strong sauce of Garlic. Garlic breedeth whelks and wounds in the body, if it be laid thereto. And if choleric men eat too much thereof, it is cause of madness and of phrensy, and grieveth the sight, and maketh it dim. Therein is virtue to put out venom, and all venomous things. Therefore it was not without cause called Treacle of churls. It helpeth best against the biting and venom of a wood hound, if it be eaten with salt and nuts, and with rue. Smape [*i.e.*, pound or crush ; Lat. *contero*] these four together, and give oft thereof to the patient, in the quantity of a great nut, and that with wine, and lay the same confection to the sore without, for it helpeth the wound, and draweth out venom, and wasteth it, and keepeth and saveth and delivereth of peril as effectually as treacle. Also it helpeth against the biting of an adder, if it be stamped and laid thereto with oil of bay.

Bartholomew (*Berthelet*), bk. xvii. § 11.

WITH fig-leaves and cummin it is laid on against the bitings of the mouse called a shrew.

Gerard's "Herbal," *s.v.*

GARLIC has so strong a scent that the leopard not being able to endure it, runs away. So that if any one rubs garlic on any place, the leopard springs away, and does not stay. It drives away serpents and scorpions by its smell.

Hortus Sanitatis, bk. i. § 14.

GARLIC being sown when the moon is under the earth, and plucked up when the moon is above the earth, it is said that then his stinking smell will be gone. Garlic will be made the sweeter, if in the planting thereof, you do set the stones of olives round about it. Or else if you set the garlic bruised.

Lupton, "A Thousand Notable Things," bk. vii. § 80.

COCKS that eat Garlic are most stout to fight; therefore travellers do often bite thereof, and also such as follow wars; because it increaseth agility, strengtheneth them, and makes them bold. It is given to horses with bread and wine, at the hour of battle or conflict, to make them more fierce, lively, and to suffer more easily their labour and travail.

Ibid., bk. viii. § 79.

Gem.

KING HENRY VIII., ii. 3, 78.

GEM hath that name for it shineth as gum. Of precious stones some breed in bodies of fowls and of creeping beasts. But from whence-so-ever precious stones come they be found endowed by the grace of God with passing great virtue when they be noble and very [*i.e.*, genuine].

Bartholomew (*Berthelet*), bk. xvi. § 48.

Gilliflower.

The year growing ancient,
Not yet on summer's death, nor on the birth
Of trembling winter, the fairest flowers o' the season
Are our carnations and streak'd gillyvors,
Which some call nature's bastards: of that kind
Our rustic garden's barren; and I care not
To get slips of them. For I have heard it said
There is an art which in their piedness shares
With great creating nature.

WINTER'S TALE, iv. 3, 79, etc.

[The Gilliflower or Gillyvor is the pink or carnation, and is to be distinguished from the Gillyflower of the wall, *i.e.*, wall-flower.]

THE Gilliflower also, the skilful do know,
Doth look to be covered in frost and in snow:
The knot and the border, and rosemary gay,
Do crave the like succour for dying away.

Tusser, "Five Hundred Points of Good Husbandry," ch. xxii. st. 22: December's Husbandry.

Ginger.

TWELFTH NIGHT, ii. 3, 126.
MEASURE FOR MEASURE, iv. 3, 6, 9.

SOME Ginger is tame, and some is wild; the wild Ginger hath more sharper savour than hath the tame, and is more sadder and faster, and not so white, but it breaketh more sooner. And the more whiter it is, and the more new, the more sharp it is and the more better. And Ginger is kept three years in good might and virtue, but afterward it waxeth dry, and worms eat and gnaw, and make holes therein and rotteth also for moisture thereof. Who that purposeth to keep Ginger by long continuance of time shall put Ginger among pepper, that the moisture of the Ginger may be tempered and suaged by dryness of the pepper.

Bartholomew (*Berthelet*), bk. xvii. § 195.

THERE are some who season Ginger with honey and some with rob [a barbarous word signifying the juice of herbs or fruits defoccate—*Cooper's* "Thesaurus"]—and some with water and salt; and that, lest it putrefy; and it is convenient in food, and is eaten with fish and salt.

Hortus Sanitatis, bk. i. § 525.

GREEN Ginger will cure me of a grievous fit of the colic.

Beaumont and Fletcher, "Scornful Lady," iv. 1.

[For a race of Ginger ("Winter's Tale," iv. 3, 20), or raze (i. "King Henry IV.," ii. 1, 26), which was probably cheap, as the Clown in the "Winter's Tale" says that he may beg it, compare Greene's "Looking-Glass for London and England":

"I spent eleven pence [for ale], beside three races of Ginger. Tapster, ho! for the king a cup of ale and a fresh toast; here's two races more." Race was probably a definite quantity, as the Clown says ("Winter's Tale," *loc. cit.*), "a race or two of ginger"; so also in i. "King Henry IV.," ii. 1, 26, "two razes of ginger"; and in "The Life and Death of Thomas Lord Cromwell" (1602 or 1613) the phrase occurs.]

THAT cinnamon, Ginger, clove, mace, and nutmeg are but the several parts and fruits of the same tree, is the common belief of those which daily use them.

Sir Thos. Browne, "Vulgar Errors," bk. ii. ch. vi.

Glow-worm.

HAMLET, i. 5, 89, 90.

THE Glow-worm is a little beast, with feet and with wings, and is therefore sometime accounted among volatiles, and he shineth in darkness as a candle, and namely about the hinder parts, and is foul and dark in full light. And infecteth and smiteth his hand that him toucheth. And though he be unseen in light, yet he fleëth light, and hateth it, and goeth only by night.

Bartholomew (*Berthelet*), bk. xviii. § 77.

CERTAIN worms that shine in the night called Glow-worms, being well stopped in a glass, and covered within hot horse-dung, standing there a certain time, will be resolved into a liquor, which being mixed with like proportion of quicksilver, first cleansed and purged: which will be within half-a-dozen times washing in pure vinegar, mixed with bay-salt, which after every washing and rubbing must be cast away, and then hot water put to the quicksilver, and therewith washed, and then put and closed in a fair, bright and pure glass, and so hanged up in the midst of a house, or other place or room: will give such a light in the dark, as the moon doth, when she shines in a bright night. *Lupton*, "A Thousand Notable Things," bk. iv. § 40.

To make a light that never shall fail.—Take the worms that shine in the night called Glow-worms, stamp them, and let them stand till the shining matter be above, then with a feather take of the same shining matter, and mingle it with quicksilver, and so put it into a vial and hang the same in a dark place, and it will give light. This I had out of an old book, which is not much unlike the description of Mizaldus.

Lupton, "A Thousand Notable Things," bk. viii. § 84.

WHERE the Glow-worm creepeth in the night, no adder will go in the day. *Lilly*, "Campaspe," Epilogue.

[Albertus Magnus ("Of the Wonders of the World") states that you may make a carbuncle of Glow-worms, treated according to the directions given in the first quotation from *Lupton's* "Notable Things."]

Gnat.

COMEDY OF ERRORS, ii. 2, 30.

A GNAT is a little fly, and is accounted among volatiles, as the bee is, though he have the body of a worm with many feet. And is gendered of rotted or corrupt vapours of carrions and corrupt place of marais [*i.e.*, marshes]. By continual flapping of wings he maketh noise in the air, as though he [w]hurred; and sitteth gladly upon carrions, botches, scabs and sores; and is full noyful to scabbed horses and sore-backed, and grieveth sleeping men with noise and with biting, and waketh them of their rest, and fleëth about most by night, and pierceth and biteth members upon the which he sitteth, and draweth toward light, and so unwarily he falleth into a candle or into the fire. And for covetous[ness] for to see light, he burneth himself oft, and is best to feeding of swallows.

Bartholomew (*Berthelet*), bk. xii. § 12.

IF any list to sleep, and lay by him the branches of moist hemp, Gnats will not trouble him nor come nigh him.

Lupton, "A Thousand Notable Things," bk. iv. § 47.

THE network coverlid spread on beds, we at this day name a canopy, a thing to catch all manner of Gnats. The Gnat seems to be a kind of fly, yet as flies love sweet things, Gnats love things sour and tart; the flies do couple, the Gnats do not. By their goodwill, they will wound none but the fairest. Gnats seem to be more worthy esteem than the ordinary sort of almanac-makers; for they will tell you the weather at all times, and for nothing, and that more certainly and truly. For if the Gnats near sunset do play up and down in open air, they presage heat; if in the shade, warm and mild showers; but if they altogether sting those that pass by them, then expect cold weather and very much rain. When a Gnat comes forth of the oak-apple about Michaelmas, it foretells war and hostility; if a spider, dearth; if a worm, fertility and fruitfulness. If any one would find water either in a hill or valley, let him observe the sun-rising, and where the Gnats whirl round in form of an obelisk, underneath there is water to be found. Yea, dreams of Gnats do foretell news of war or a disease, and that so much the more dangerous as it shall be apprehended to approach the more principal parts of the body. Hang some horse-hair and make it fast in the middle of the doors, and Gnats will not come in at it. Our countrymen that live about the fens have invented a fen-canopy, being made of a broad, plain, half-dry, somewhat hard piece or many pieces together of cow's dung, and these they hang at their bed's feet; with the smell and juice whereof the Gnats being very much taken, and feeding thereon all the night long, let them sleep quietly in their beds.

Mouffet, "Theatre of Insects," bk. i. ch. xiii.

Goat.

i. KING HENRY IV., iii. 1, 39.

THE Goat breathes at the ears, and not at the nose, and is seld without fever. And we mean not that Goats see less by night than by day. And if a man draw one out of the flock by the beard, the others be astonied and behold. And also the same happeth when one of them biteth a certain herb. And if the Goat conceive afore the northern wind, she yeaneth males, and if she conceive afore the southern wind, she yeaneth females. And if a man take a Goat, and rear him up suddenly, then the other rear them also, and behold him sadly. Serpents be chased and driven away with ashes of Goats' horns, and with their wool burnt.

And by remedy of Goats' horns divers manner kind of venom is overcome. With new Goat-skins wounds be holp and healed. Goat's blood meddled with marrow and sod excludeth poison of venom, biting of creeping worms, and smiting of scorpions be saved and healed. And a certain beast sucketh goat's milk of the udder and teats, and then the milk is destroyed and wasted, and the goat waxeth blind thereby. *Bartholomew* (*Berthelet*), bk. xviii. § 24.

WILD Goats dwell in high rocks and crags; and if they perceive sometime that they be pursued of men or of wild beasts, they fall down headlong out of the high crags, and save themselves harmless on their own horns. Also the leopard drinketh milk of the wild Goat, and voideth sorrow and woe. *Ibid.*, § 22.

WHEN he is wounded, he eateth Dragon tea, and taketh so the arrow out of the body. Serpents hate and flee the wild Goat, and may not suffer the breath of him.

Ibid., § 35.

THE hot blood of the Goat buck nesheth and carveth the hard adamant stone, that neither fire nor iron may overcome. And the Goat buck hath many and strong horns, and much fatness and namely within about the reins, and then he dieth lightly, but the fatness be withdrawn. Sometime it happed that a Goat buck was seen with horns in the legs, and that was full wonderful to see. The liver of the Goat buck helpeth against biting of the wood hound. *Ibid.*, § 60.

GOATS will not stray nor wander, if you cut off their beards.

Lupton's "A Thousand Notable Things," bk. v. § 58.

THE bite of the Goat is deadly to trees. And Goats die if they lick honey. They live on venomous herbs. If Goats drink or eat out of vessels of tamarisk, they will have no spleen. If they lick serpents after these have cast their skin, they will not grow old, though they become white. The Goat does not see well in the day-light; but its sight is more acute by night. The eyes of the Goat shine by night, and they throw out light. Also he-Goats have more teeth than she-Goats. Goat's cheese appeases all wounds and pains if laid upon them. Their hoofs burnt and pounded with liquid pitch cure baldness. Their blood does as much; and if it be drunk destroys venom.

Hortus Sanitatis, bk. ii. § 22.

UPON provocation the he-Goat striketh through an ordinary piece of armour or shield at one blow,—his force and the sharpness of his horns are so pregnable. Goats foresee and foreshew change of weather, for they depart from their stables, and run wantonly abroad before showers, and afterward, having well fed, of their own accord return to their folds again. Goats take breath through their ears; and certain Goats have a certain hole or passage in the middle of their head, betwixt the horns, which goeth directly unto the liver, and the same stopped with liquid wax suffocateth or stifleth the beast. There is no beast that heareth so perfectly and so sure as a Goat, for he is not only holp in this sense with his ears, but also hath the organ of hearing in part of his throat. With Goat's milk wine is preserved from corruption by sourness. Of the suet and fat of Goats are the best candles made, because it is hard and not over liquid. The blood of a Goat scoureth rusty iron better than a file. The loadstone draweth iron, and the same, being rubbed with garlic, dieth and loseth that property, but being dipped again in Goat's blood, reviveth and recovereth the former nature. In ancient time they made fruitful their vineyards by this means:—they took three horns of a female Goat, and buried them in the earth with their points or tops downward to the root of the vine stocks. The gall of a female Goat put into a vessel and set in the earth hath a natural power to draw Goats unto it. Herein appeareth the pride of this beast, that he scorneth to come behind either cattle or sheep, but always goeth before. Goats love singularity, and may well be called schismatics among cattle; in great stocks they are soon infected with the pestilence. The wild Goats of Crete eat dittany against the strokes of darts; and Goats by licking the leaves of tamarisk lose their gall. The rhododendron is poison to Goats, and yet the same helpeth a man against the venom of serpents. Also they avoid cummin, for it maketh them mad, or bringeth upon them lethargies, and such like infirmities. He avoideth also the spittle of man, for it is hurtful to him, and yet he eateth many venomous herbs and groweth fat thereby. The Goats of Cephalonia drink not every day like other Goats, but only once or twice in six months. And wheras all other kind of cattle, when they are sick, consume and pule away by little and little, only Goats perish suddenly, insomuch as all that are sick are unrecoverable ; and the other of the stock must be instantly let blood and separated before the infection overspread all. The female Goat easeth the pain of her eyes by pricking them upon a bullrush, and the male Goat by pricking them upon a thorn. The females never wink in their sleep, being herein like the roe-bucks. There are certain birds called [Goatsuckers] because of their sucking of Goats, and when these have sucked a Goat, she presently falleth blind. Young wild Goats gather meat and bring it to their mothers in their age, and likewise they run to the rivers or watering-places, and with their mouths suck up water, which they bring to quench the thirst of their parents; and whereas their bodies are rough and ugly to look upon, the young ones lick them over with their tongues, making them smooth and neat. The horns [of the wild Goats] serve them [the shepherds] instead of buckets to draw water out of the running streams ; they are so great, that no man is able to drink them off at one draught. The wild Goats of Egypt are said never to be hurt by scorpions.

Topsell, "Four-footed Beasts," pp. 181-94.

If Goat's blood be taken warm, with vinegar and the juice of hay and the like be boiled with glass, it makes the glass soft like paste, and it may be thrown against a wall, and will not break, and if the aforesaid be poured into a vase, and the face anointed with it, strange and horrible things will appear, and the man will think that he must die.

Albertus Magnus, "Of the Virtues of Animals."

Goose, Gosling.

LOVE'S LABOUR'S LOST, iii. 1, 102, etc.
CORIOLANUS, v. 3, 35.

IN the Alps there is a kind of Goose, biggest of all birds except the ostrich; but so heavy that it may be taken immovable on the ground by the hand. There is no animal which so quickly perceives the scent of man as the Goose. Its fat helps against baldness.

Hortus Sanitatis, bk. iii. § 10.

IF a man steal their eggs from them, they lay still, and never give over till they be ready to burst with laying. If one of their Goslings be stung never so little by a nettle, it will die of it. Their greedy feeding also is their bane, for one while they will eat till they burst again; another whiles kill themselves with straining their own selves; for if they chance to catch hold of a root with their bill, they will bite and pull so hard for to have it, that many times they break their own necks withal, before they leave their hold. Against the stinging of nettles the remedy is, that so soon as they be hatched there be some nettle roots laid under their bed of straw. *Holland's Pliny*, bk. x. ch. lix.

IT is said that all summer long even unto the fall of the leaf, Geese and ravens be continually sick.

Ibid., bk. xxix. ch. iii.

[WHITBY.] It is also ascribed to the sanctity of Hilda, that those wild Geese (which in winter fly in great flocks to the unfrozen lakes and rivers in the southern parts), to the great amazement of every body, fall down suddenly upon the ground, when they are in their flight over certain neighbouring fields hereabouts; a relation that I should not have given, if I had not received it from several very credible persons. *Camden's* "Britannia," col. 906-7 (ed. 1722).

AN excellent pickled Goose, a new service.

Dekker and Webster, "Westward Ho!" i. 2.

YOUNG gentlemen shall be eaten up (for dainty meat) as if they were pickled Geese, or baked woodcocks.

Dekker, "Raven's Almanack."

JOHN DE LA HAY held a parcel of land of Will. Barneby Lord of Lastres in County Hereford, and was to render thence 20d. yearly, and one Goose fit for the Lord's dinner on the Feast of St. Michael the Archangel.

Blount's "Jocular Tenures," 10 Edward IV., p. 8.

[An old proverb: "Teach him a trick to shoe the goose"—("Bacchus' Bounty," and "Parliament of Birds").]

SOME hold an opinion that in over rank soils Goose-dung doth so qualify the batableness [fertility] of the soil, that their cattle is thereby kept from the garget, and sundry other diseases, although some of them come to their ends now and then by licking up of their feathers.

Holinshed, "Description of England," p. 222.

Gooseberry.

ii. KING HENRY IV., i. 2, 194.

GOOSEBERRY, because they commonly make goose-sauce with Gooseberries. Its flavour is like that of the green fig.

Minsheu's Dictionary, *s.v.*

IN English Gooseberry, and fea-berry in Cheshire. The fruit is used in divers sauces for meat, as those that are skilfull in cookery can better tell than myself. They are used in broths instead of verjuice, which maketh the broth not only pleasant to the taste, but is greatly profitable to such as are troubled with a hot burning ague. They provoke appetite. [The fruit must have been very small, as] there is another whose fruit is almost as big as a small cherry. These plants do grow in our London gardens and elsewhere in great abundance.

Gerard's "Herbal," bk. iii. ch. xxii.

[Johnson (*Gerard's* "Herbal," *loc. cit.*) mentions six sorts of Gooseberries: the long green, the great yellowish, the blue, the great round red, the long red, and the prickly.]

Grape.

ALL'S WELL THAT ENDS WELL, ii. 1, 73.
AS YOU LIKE IT, v. 1, 39.
MIDSUMMER NIGHT'S DREAM, iii. 1, 170.

THERE is a kind of black Grape named *Inerticula*, as a man would say dull and harmless; but they that so called it might more justly have named it The sober Grape; the wine made thereof is very commendable when it is old, howbeit nothing hurtful, for never makes it any man drunk; and this property hath it alone by itself.

Holland's Pliny, bk. xiv. chs. ii., iii.

GRAPES may be kept the whole year. Take the meal of mustard-seed, and strew in the bottom of any earthen pot well leaded; whereupon you shall lay the fairest bunches of the ripest Grapes, the which you shall cover with more of the foresaid meal, and lay upon that another sort of Grapes, so doing until the pot be full. Then shall you fill up the pot to the brim with a kind of sweet wine called must. The pot being very close covered shall be set into some cellar or other cold place. The Grapes you may take forth at your pleasure, washing them with fair water from the powder.

Gerard's "Herbal," bk. ii. ch. cccxxiii.

Gorse.

TEMPEST, iv. 1, 180.

[Distinguished from "furze."]

V. Furze.

Grass.

GRASS cometh of the green, and is pleasing in sight, and liking to beasts in pasture and meat, and comforteth the sick in doing, for as in roots so in herbs and Grass be many manner virtues. Herbs and Grass love stern weather, rain and great showers, for heat and colour of herbs need much moisture. Hounds know this herb, and eat it to purge themselves, but they do it so privily, that unneath men may spy it.

Bartholomew (*Berthelet*), bk. xvii. § 76.

The earth found in or about a man or woman's skull is a singular depilatory, and fetcheth away the hair of the eyebrows. As for the Grass or weed that groweth therein (if any such may be found) it causeth the teeth to fall out of the head with chewing only.

Holland's Pliny, bk. xxviii. ch. iv. p. 302, G.

Five-leaved Grass, through Jupiter's force, doth resist venom or poison. Whereof if one leaf twice every day, morning and evening, be drunken with wine, it is said to put away the quotidian ague. Three leaves the tertian ague. And four leaves the quartan ague.

Lupton, "A Thousand Notable Things," bk. iii. § 45.

The fine Grass which groweth upon the banks of [the Dove] is so fine and batable [luxuriant], that there goeth a proverb upon the same; so oft as a man will commend his pasture, to say that there groweth no better feed on Dove-bank. *Holinshed*, "Description of England," p. 98.

Grasshopper.

Romeo and Juliet, i. 4, 60.

They never alter their place, or at least very seldom; or if they do, they are ever after silent, they sing no more; so much doth the love of their native soil prevail with them. If clay be not dug up in due time, it will breed Grasshoppers. The Grasshoppers abounding in the end of spring do foretell a sickly year to come. Oftentimes their coming and singing doth portend the happy state of things. What year that few of them are to be seen, they presage dearness of victuals, and scarcity of all things else. *Mouffet*, "Theatre of Insects," bk. i. ch. xvii.

V. Locust.

Griffin.

Midsummer Night's Dream, ii. 1, 232.

A Gripe is accounted among volatiles. And the Gripe is four-footed, and like to the eagle in head and in wings; and is like to the lion in the other part of the body, and dwelleth in those hills that be called Hyperborean, and be most enemies to horses and men, and grieveth them most, and layeth in his nest a stone that hight Smaragdus against venomous beasts of the mountain.

Bartholomew (*Berthelet*), bk. xii. § 19.

THE Gripes are of colour of a dark ochre on the base, their breast of purple colour, their wings brown and white, their talons black, and the beak turning as doth the eagle's; he is more higher than the lion,—the hinder feet cloven as the stag's,—able to carry away the weight of two men, a stag, or the like beast.

Batman's addition to *Bartholomew*, bk. xii. § 19.

GRIFFINS dig up gold and delight in looking at it when dug up. The body of a large Griffin is larger than eight lions of those parts; for having killed an ox, a horse, or even an armed man, it lifts them up and carries them off in its flight. *Hortus Sanitatis*, bk. iii. § 56.

THEY build their nests of the gold which they dig up, and lay two eggs larger, harder, hotter and drier than those of eagles.

Jonston, "Natural History," bk. iii., appendix, ch. i.

GRIPES keep the mountains, in the which be gems and precious stones, as emerald and jasper, and suffer them not to be taken from thence. And in some countries in Scythia is plenty of gold and of precious stones; but for great Gripes men dare not come thither openly, but seld for fierceness of Gripes. There is best emerald and crystal. And the Gripe hath so great claws and so large, that of them be made cups that be set upon boards of kings.

Bartholomew (*Berthelet*), bk. xviii. § 56.

IN that country [Bacharie] be many Griffins, more plenty than in any other country. One Griffin is more great and stronger than an hundred eagles such as we have among us. For one Griffin there will bear, flying to his nest, a great horse, or two oxen yoked together, as they go at the plough. For he hath his talons so long and so large and great upon his feet, as though they were horns of great oxen, or of bugles [*i.e.*, buffaloes] or of kine; so that men make cups of them to drink of; and of their ribs and of the pens [*i.e.*, feathers] of their wings, men make bows full strong, to shoot with arrows and quarrels.

Sir John Mandeville, ch. xxvi. *ad fin.*

GRIPES make their nests of gold, though their coats are feathers. *Lilly*, "Galatea," ii. 3.

THE Griffin never spreadeth her wings in the sun, when she hath any sick feathers.

Lilly, "Sappho and Phaon" (Prologue at the Blackfriars).

Gudgeon.

This fool gudgeon.

MERCHANT OF VENICE, i. 1, 102.

SOME say that the Gudgeon feeds on dead carcases, but fishermen hold this to be a fable.

Hortus Sanitatis, bk. iv. § 41.

SOME have said that the Gudgeon is generated from the brain of horses.

Jonston, "Natural History," bk. ii. tit. i. ch. x.

WHAT fish soever you be, you have made both me and Philantus to swallow a Gudgeon.

"Euphues"; so *Dekker's* "Honest Whore," part ii. ii. 2.

Gurnet.

A soused gurnet.

i. KING HENRY IV., v. 2, 11.

[GURNET is called in Latin] *cuculus*, either from the sound which it makes in common with the bird of that name [*i.e.* cuckoo], or because, when it is taken in nets, it utters the word cu.

Jonston, "Natural History," bk. ii. tit. iii. ch. i. p. 2.

[Gurnets cost 2s. 6d. to 3s. each in 1573.
Soused Gurnet was a term of reproach; *vide Steevens'* Shakespeare, vol. viii. p. 549.]

Halcyon.

St. Martin's summer, halcyon days.
i. KING HENRY VI., i. 2, 131.

Turn their halcyon beaks with every gale.
KING LEAR, ii. 2, 84.

A BIRD called also King's-fisher, because she fisheth in the sea, and casteth herself with such force at the fishes. She conceiveth in the sea, and in it she brings forth her young—and that in chill and cold weather; and meanwhile the heaven is serene, and the sea tranquil, nor agitated by troubles of winds. Hence those serene days are called Halcyon-days.

Minsheu's Dictionary, *s.v.*

SHE deposits her eggs in the sand, and that in mid-winter, when the sea rises highest, and the waves beat very strongly on the shore; but while she hatches out, the sea grows suddenly quiet, and all windy storms cease. And she sits on her eggs for seven days, and then brings out her young, whom she rears for other seven days. And therefore seamen watch for these xiv days, expecting calms. Her nest cannot be cut by iron, but is broken by a strong knock.

Hortus Sanitatis, bk. iii. § 8.

THERE is a second kind of them breeding about the sea-side, differing both in quantity and also in voice; for it singeth not as the former do, which are lesser; for they haunt rivers, and sing among the flags and reeds. It is a very great chance to see one of these Halcyons, and never are they seen but about the setting of the star *Virgiliæ* (*i.e.*, the Brood-hen); or else near mid-snmmer or mid-winter: for otherwhiles they will fly about a ship, but soon are they gone again and hidden. In the beginning of December they build. Their nests are wondrously made, in fashion of a round ball; the mouth or entry thereof standeth somewhat out, and is very narrow, much like unto great sponges. And no man could ever find of what they be made. Some think they are framed of the sharp-pointed pricks of some fishes, for of fish these birds live.

Holland's Pliny, bk. x. ch. xxxii.

THERE is engendered in the sea also that which is called *Halcyoneum*, made as some think of the nests of the birds Halcyons; but, as others suppose, of the filthy foam of the sea. Four kinds there be of it.

Holland's Pliny, bk. xxxii. ch. viii.

INTO the nest of an Halcyon no bird can enter but the Halcyon. *Lilly*, "Sappho and Phaon," iii. 3.

As the birds Halcyon which exceed in whiteness, I hatch young ones that exceed in blackness.

"Euphues' Golden Legacie."

A LITTLE bird called the King's fisher, being hanged up in the air by the neck, his neb or bill will be always direct or straight against the wind. This was told me for a very truth by one that knew it by proof, as he said.

Lupton, "A Thousand Notable Things," bk. x. § 96.

BUT now how stands the wind?
Into what corner peers my Halcyon's bill?
Ha! To the East? Yes! See, how stand the vanes?
East and by south. *Marlowe*, "Jew of Malta," i. 1.

As a Halcyon with her turning breast
Demonstrates wind from wind, and east from west.
Storer, "Life and Death of Thomas Wolsey, Cardinal" (1599).

Hare.

[Melancholy] hare.
i. KING HENRY IV., i. 2, 86.

THE Hare is fearful, and fighteth not, and is feeble of sight, as other beasts be, that close not the eye-lids in sleeping; and is better of hearing than of sight, namely when he reareth up the ears. His ears be full long and pliant, and that is needful for to defend the eyes that be open, and not defended with covering, nor with heling to keep them from gnats and flies great and small for against noyful [*i.e.*, noxious] things, kind giveth remedy to creatures. *Bartholomew* (*Berthelet*), bk. xviii. § 68.

Like your melancholy Hare
Feed after midnight.

Webster, "Vittoria Corombona."

HOWSOEVER Hares are thought to nourish melancholy, yet they are eaten as venison, both roasted and boiled.

Fynes Moryson, "Itinerary," part iii. p. 149.

[Hares were roasted (second part of "The Good Huswife's Jewel," 1597, p. 66) with parsley, thyme, savory, cream, butter, small raisins and barberries worked all together in the Hare's belly, and served with venison sauce.]

THE juice of henbane, mixed with the blood of an Hare, and sod within the skin of a Hare, it is said that all the Hares will gather together, which be within that trace where it is buried. This was affirmed for truth to Mizaldus.

Lupton, "A Thousand Notable Things," bk. ii. § 5.

[HARE-LIP comes of seeing a Hare or longing for its flesh. *Ibid.*, bk. ii. § 6.]

THE blood of an Hare dried and made in powder, and thrown upon flesh newly roasted or sodden, makes the same flesh seem to be bloody and corrupt. So that they that be present, and sees the same, unless such as know the secret thereof, will loathe to eat thereof (Mizaldus).

Ibid., bk. vii. § 66.

WITH its brain boys' gums are cleansed; for it has the property to make the teeth come quickly, and without pain. Its head burnt with bear's grease, and used as a plaster, helps baldness. *Hortus Sanitatis*, bk. ii. § 83.

IN Chersonesus all the Hares have ordinarily two livers; and (a wondrous thing it is to tell) if they be brought into other countries, one of the said livers they lose.

Holland's Pliny, bk. xi. p. 341.

SOME creatures there are that will never be fat, as the Hare and partridge. *Ibid.*, p. 344.

THE hairiest creature of all other is the Hare.

Holland's Pliny, bk. xi. p. 347.

MEN have assayed to make cloth of Hares' and Cony's hair; but in the hand they are not so soft as is the fur upon the skin or case; neither will they last, by reason that the hair is short, and will soon shed.

Ibid., p. 232.

THE common sort of people are persuaded, that the meat of this kind of venison [*i.e.*, Hare's flesh] causeth them that feed upon it to look fair, lovely and gracious, for a week together afterwards. There must needs be some cause and reason of this settled opinion, which hath thus generally carried the world away to think so.

Ibid., bk. xxviii. p. 341.

THE eye-lids coming from the brows are too short to cover their eyes, and therefore this sense is very weak in them; and besides their over-much sleep, their fear of dogs and swiftness causeth them to see the less; when they watch, they shut their eyes, and when they sleep they open them. The common sort of people suppose they are one year male, and another female. Men find in Hares certain little bladders filled with matter, and against rain Hares suck thereout a certain humour, and anoint their bodies all over therewith, and so are defended in time of rain. Hares never drink, but content themselves with the dew, and for that cause they often fall rotten. She keepeth not her young ones together in one litter, but layeth them a furlong one from another, that so she may not lose them all together, if peradventure men or beasts light upon them. The ears of this beast are like angels' wings, ships' sails and rowing oars, to help her in her flight. The eating of Hares procureth sleep. A waistcoat made of Hares' skins straightens the bodies of young and old. The rennet being mingled with vinegar is drunk against poison; and also if a man or beast be anointed with it, no serpent, scorpion, spider or wild mouse, whose teeth are venomous, will venture to sting the body so anointed. The same being mingled with snails or any other shell-fish, which feed upon green herbs or leaves, draweth forth thorns, darts, arrows or reeds out of the belly.

Topsell, "Four-footed Beasts," pp. 208-16.

THE feet of a Hare together with the stone otherwise the head of an ousel move a man to boldness, so that he fears not death. And if it be bound on the left arm, he will go whither he will, and return safely without danger. And if it be given to a dog to eat with the heart of a weasel, he will make no noise from thenceforth, even if he is being killed.

Albertus Magnus, "Of Virtues of Animals."

WITH [the red deer] in degree of venerie are accounted the Hare, boar and wolf. As for Hares, they run at their own adventure, except some gentleman or other (for his pleasure) do make an enclosure for them.

Holinshed, "Description of England," p. 226.

V. Cony.

Hare-bell.

CYMBELINE, iv. 2, 222.

THE roots, being beaten and applied with white wine, hinder or keep back the growth of hairs. The root boiled in wine and drunk helpeth against the venomous bitings of the field-spider. The seed is of the same virtue.

Gerard's "Herbal," *s.v.*

Harlock.

KING LEAR, iv. 4, 4.

[A doubtful reading. Hardock is *Arctium Lappa*. "Harlock" is used by Drayton, but the plant has not been identified. It is possible that "charlock" or "burdock" may be the right word.]

Hart.

AS YOU LIKE IT, iii. 2, 107.

A HART is a stag of five years old complete. And if the King or Queen do hunt him, and he escape away alive, then afterward he is called a Hart Royal. And if the beast, by the King's or Queen's hunting, be chased out of the forest, and so escape, proclamation is commonly made in the places thereabout, that in regard of the pastime that the beast hath showed to the King or Queen, none shall hurt him or hinder him from returning to the forest; and then is he a Hart Royal Proclaimed.

Minsheu's Dictionary, *s.v.*

Harts be enemies to serpents; which when they feel themselves grieved with sickness, they draw them with breath of their nostrils out of their dens, and, the malice of the venom overcome, they are repaired with feeding of them. And they taught first the virtue of the herb Dittany, for they eat thereof, and cast out arrows and arrow-heads, when they be wounded of hunters. And they wonder of noise of pipes, and have liking in accord of melody, and they hear well when they rear up their ears, and bear down the ears when they swim and pass rivers and great waters. And then in swimming the stronger swim tofore, and the feebler lay their heads upon the loins of the stronger. And the Hart is most pleasing beast, and runneth wilfully and fleëth to a man when he is overset with hounds. [And in rutting time] the males wax cruel, and dig up clods and stones with their feet, and then their snouts be black until they be washed with rain. [And after the female has calved] the male eateth busily; and when he feeleth himself too fat, he seeketh dens and lurking-places, for he dreadeth damage and harm by heaviness of body. And when the Hart casteth his right horn, for envy he hideth it, and is sorry if any man hath medicine thereof. Serpents flee and avoid the odour and smell of burning of an Hart's horn. His rennet is good against all biting of serpents. Also the Hart's blood and hare's blood congealeth never. And the Hart roareth, cryeth and weepeth when he is taken. And when the Hind feeleth heaviness, she swalloweth a stone, and is holpen by virtue of that stone. *Bartholomew* (*Berthelet*), bk. xviii. § 30.

Give the bone of a Hart's heart, ground, to a barren woman in drink, and thou shalt see the glory of God.

Batman's addition to *Bartholomew*, bk. xviii. § 30.

HARTS being the most cowardly and heartless creatures have also the largest horns.

Dekker, "News from Hell."

OXEN, kine, bullocks or horses shall not be troubled with any disease, if you hang a Hart's horn upon them.

Lupton, "A Thousand Notable Things," bk. vi. § 53.

A HART doth so abhor a ram, that he cannot abide the sight of him. *Ibid.*, bk. ix. § 35.

CERTAIN worms are bred in the bowels or guts of the Hart, and they are destroyed by the eating of serpents, which the Hart doth allure with the breath of his nose to come out of their hole or den; and lest the poison of them should hurt him, he goes apace to some fair spring of water, and whiles all his whole body is therein unto the lips, little drops or tears distil out of his eyes, which at length increaseth to a thing as big as a walnut, and are in manner of a stone, and when he perceives he hath thereby avoided all the poison, and being come forth of the water, with the rubbing of his eyes at a tree, the same lump or stone (being a hindrance to his sight) he gets away. Which matter or stone is a thing most effectual against any venom or poison. The Arabian physicians call the stone Bezoar. *Ibid.*, bk. x. § 21.

THE Hart hath a worm in his head, which vexes him constantly in the spring. But every animal and man himself has a worm under the tongue. The Hart, where he finds a serpent, fills his mouth with water, and pours it into the hole, then with the breath of his mouth he draws the serpent out, and treads on it with his feet and kills it, and eats it. Any one who is wrapped in the hide of a Hart does not fear serpents. The end of a hart's tail is venomous. *Hortus Sanitatis*, bk. ii. § 34.

HARTS being stung with a kind of spider, or some such venomous vermin, they cure themselves with eating crayfishes or fresh-water crabs.

Holland's Pliny, bk. viii. ch. xxvii.

THE stag and hind feeling themselves poisoned with some venomous weed among grass where they pasture go by and by to the artichoke, and therewith cure themselves.

Holland's Pliny, bk. viii. ch. xxvii.

THIS creature of all diseases is not subject to the fever, but he is good to cure it.

Ibid., bk. ii. ch. xxxii.

IF together with deer's blood, there be burnt the herb Dragon, Bastard Marjoram, and Orchanet, in a fire made with Lentisk wood, serpents will gather round together into an heap; take away the same blood and put into the fire the root of Pyrethrum (Pellitory of Spain), they will scatter asunder again. *Ibid.*, bk. xxviii. ch. ix.

HARTS are deceived with music, for they so love that harmony, that they forbear their food to follow it. They live very long—2,112 years. The bones of young Harts are applied for making of pipes, but if a young one be pricked in his legs with cactus, his bones will never make pipes. If men drink in pots wherein are wrought Harts' horns, it will weaken all force of venom. The magicians have also devised that if the fat of a dragon's heart be bound up in the skin of a roe, with the nerves of a Hart, it promiseth victory to him that beareth it on his shoulder, and that if the teeth be so bound in a roe's skin, it maketh one's lord, master, or all superior powers, exorable and appeased towards their husbands and suitors. Orpheus, in his Book of Stones, commandeth a husband to carry about him a Hart's horn, if he will live in amity and concord with his wife.

Topsell, "Four-footed Beasts," pp. 101-5.

THE young males which our fallow deer do bring forth are commonly named according to their several ages;—for the first year it is a fawn, the second a pricket, the third a sorrel, the fourth a sore, the fifth a buck of the first head. In examining the condition of our red deer, I find that the young male is called in the first year a calf, in the second a brocket, the third a spay, the fourth a stagon or stag, the fifth a great stag, the sixth an Hart, and so forth unto his death. And with him in degree of venery are accounted the hare, boar and wolf. Of these also the stag is accounted for the most noble game, the fallow deer is the next, then the roe, whereof we have indifferent store, and last of all the hare.

Holinshed, "Description of Britain," p. 226.

Hawk.

TAMING OF THE SHREW, Induction, Sc. 2, 45.

WE have the eagle; the lanner [male] and the lanneret [female]; the tiercel and the goshawk; the musket [male-sparrow-hawk] and the sparhawk; the jack and the hobby [a small Hawk]; and finally some though very few marlions [merlins]. And these are all the Hawks that I do hear as yet to be bred within this island. Howbeit as these are not wanting with us, so are they not very plentiful; wherefore such as delight in Hawking do make their chief purveyance and provision for the same out of Dansk [Denmark], Germany and the East countries; from whence we have them in great abundance, and at excessive prices, whereas at home they are sold for almost right naught. The sparhawk is enemy to young children, as is also the ape; but of the peacock she is marvellously afraid, and so appalled, that all courage and stomach for a time is taken from her upon the sight thereof.

Holinshed, "Description of England," p. 227; ch. v.

THE goshawk is a royal bird, and is armed more with boldness than with claws, and as much as kind taketh from her in quantity of body, he rewardeth her with boldness of heart. And she is a covetous fowl to take other fowls. Also such Hawks be cruel against their birds, so that they take from them meat when they be fledge and ripe, and they beat and drive them out of their nest, as the eagle doth her birds. And some such Hawks be thieves of the air only, and some of the earth only. And the more sharp her breast is, the better she is of flight. And the goshawk hath this property, that in age, when she feeleth herself grieved with heaviness and weight of feathers, she spreadeth her wings against the beams of the sun, when the wind is south, and so by sudden weather and resolving heat the pores be opened. And when the pores be so opened, she smiteth and flappeth her wings, and in so doing the old feathers leap out, and new grow; and so the new feathers make her in better state and the more able to flight. And two kinds there be of such fowls: for some be tame, and some be wild. And he that is tame taketh wild fowls, and taketh them to his lord; and he that is wild taketh tame fowls. And this goshawk is of a disdainous kind; for if she fail by any hap of the prey that she reseth to, that day uneath she cometh to her lord's hand. And they be borne on the left hand, that they may somewhat take of the right hand, and be fed therewith.

Bartholomew (Berthelet), bk. xii. § 2., where also are various directions for keeping and feeding hawks, for which see also *Markham's* "Husbandry," etc.

WE find in falconry 16 kinds of Hawks or fowls that prey. Of which the Circos (which is lame and limpeth of one leg) was held in ancient time for the luckiest augury in case of weddings and of cattle. In general, Hawks are divided into sundry and distinct kinds by their greediness more or less. *Holland's Pliny*, bk. x. ch. viii.

THE Hawk holds beneath its talons all night a bird that fortune offers it at night-time, but when the sun rises the Hawk even though hungry lets the bird fly away, and if he meets it at some other time, does not pursue it.

Hortus Sanitatis, bk. iii. § 4.

Hazel.

TAMING OF THE SHREW, ii. 1, 255.

V. **Filbert.**

Heart's-ease.

ROMEO AND JULIET, iv. 5 (not of the plant).

[A writer in the *Saturday Review* (March 24, 1894) says that "Heart's-ease" is properly the name of the wall-flower, but he gives no authority for the statement.]

V. **Pansy.**

Heath.

TEMPEST, i. 1, 70.

THE tender tops and flowers are good to be laid upon the bitings and stingings of any venomous beast; of these flowers the bees do gather bad honey.

Gerard's "Herbal," *s.v.*

THE leaf of this plant is an enemy to serpents.

Holland's Pliny, bk. xxiv. ch. ix.

IF it be eaten alone, it induces head-ache, therefore it should be eaten with lettuce or endive. If mixed with milk or vinegar and lozenges made of it, it can keep flesh from putrefaction. *Hortus Sanitatis*, bk. i. § 176.

A KIND of broom, whereof brushes be made.

Minsheu's Dictionary, *s.v.*

Hedgehog.

MIDSUMMER NIGHT'S DREAM, ii. 2, 9.

THE Urchin is a beast heled with pricks, hard and sharp, and his skin is closed about with pikes and pricks, and he closeth himself therewith. And is a beast of purveyance; for he climbeth upon a vine or an apple-tree, and shaketh down grapes and apples. And when they be felled, he walloweth on them, and sticketh his pricks in them, and so beareth meat to his children in that manner wise. And there is a manner kind of Urchins with a white shell and white pikes, and layeth many eggs. Also the urchin hath feeble hearing, more feeble than other beasts with hard shells, and that go on four feet. In Urchins is wit and knowing of coming of winds north or south; for he maketh a den in the ground when he is ware that such winds come. And so sometime was one in Constantinople, that had an Urchin, and knew and warned thereby that winds should come, and of what side, and none of his neighbours wist whereby he had such knowledge and warning. Also the Urchin breedeth five eggs better than other, and the eggs of some be much and great, and some be less; for some be better to seething and to defying [*i.e.*, digesting] than other. Also Urchins have a little body and many pikes, that occupy more place than the body; and the cause of many great pricks, and the littleness of the body is for feeding of the body passeth into nourishing and growing of pikes, because of scarcity of heat, and for the meat is not well defied; and therefore in his body breedeth much superfluity, and that superfluity passeth into nourishing and feeding of pricks.

Bartholomew (*Berthelet*), bk xviii. § 62.

Is a little beast with pricks, and is like to the Urchin; but he is accounted more than he. He walloweth upon apples, as the Urchin doth, which stick there on his pricks, and he beareth them into hollowness of trees. And beside the apples that he beareth on his back, alway he beareth one in his mouth. And after that he is charged with grapes or with apples, if any apple or grape fall out of the pikes in any manner wise, then for indignation he throweth away off his back all the other deal; and oft turneth again to the tree to charge him again with new charge. And his skin that is so piked is needful to men, that if there were no pikes and pricks, neshness of flesh in beasts were idle to mankind. For with such a beast's skin, cloths be cleansed and piked.

Bartholomew (*Berthelet*), bk. xviii. § 63.

THE serpent seeketh out the Hedgehog's den, and falleth upon her to kill her; the Hedgehog draweth itself up together round like a foot-ball, so that nothing appeareth on her but her thorny prickles; whereat the serpent biteth in vain, for the more she laboureth to annoy the Hedgehog, the more she is wounded and harmeth herself. The Hedge-hog rolleth upon the serpent, piercing his skin and flesh (yea, many times tearing the flesh from the bones) whereby he scapeth alive, and killeth his adversary, carrying the flesh upon his spears, like an honourable banner won from his adversary in the field. The wolf also is afraid of and flieth from the Hedge-hog; and there is a story of hatred between the hare and the Hedge-hog, for a hare was seen to pluck off the prickles from the Hedge-hog, and leave her bald, peeled and naked without any defence. With the skin, brushes are made for garments, and also it is set upon a javelin at the door to drive away dogs.

Topsell, "Four-footed Beasts," p. 219.

Hemlock.

MACBETH, iv. 1, 25.

IN English, Hemlock, Homlock, Kexe, and Herb Bennet. Hemlock is a very evil, dangerous, hurtful and poisonous herb, insomuch that whosoever taketh of it into his body dieth remediless, except the party drink some wine that is naturally hot, before the venom have taken the heart; but being drunk with wine the poison is with greater speed carried to the heart by reason whereof it killeth presently, therefore not to be applied outwardly, much less taken inwardly into the body.

Gerard's "Herbal," *s.v.*

ITS greatest strength is in the root, the second in the leaves, the least in the seed. Its leaves drive away vipers and serpents.

Hortus Sanitatis, bk. i. § 115.

Hemp.

ii. King Henry IV., ii. 1, 64.

["Hemp-seed" here, of course, refers to the use of Hemp for making ropes.]

The female Hemp [is] barren and without seed, contrary unto the nature of that sex; which is very like to the male, and one must be gathered before the other be ripe, else it will wither away, and come to no good purpose.

Gerard's "Herbal," *s.v.*

If you lay the wick of a candle to infuse or steep in the oil of Hemp-seed, and after make a tallow candle thereof, which if you do light after it be cold, the same candle will not go out with any wind, so long as the whole candle lasteth. And in like sort may lights be made to serve in the night-time, if that fine linen rags be first soaked in the oil of Hemp-seed, and dipped into molten tallow, being so bound or wrought on a staff's end, or otherwise lying in an iron or plate at the end of a staff.

Lupton, "A Thousand Notable Things," bk. x. § 23.

The juice of green Hemp-seed, being dropped into the ears, driveth out any worms or vermin there engendered, yea, and what ear-wigs or such like creatures that are gotten into them; but it will cause head-ache withal. So forcible is this plant, that if it be put into water, it will make it to gather and coagulate. *Holland's Pliny*, bk. xx. ch. xxiii.

Hen.

Short-legged hens.
ii. King Henry IV., v. 1, 28.

As some men mean if her members were meddled with gold when it is molten, the gold should waste. The Hen is a fowl of great laying, and layeth many eggs without treading, and they be called wind-eggs, and be more unsavoury and less worthy than other eggs. A Hen is a mild bird about chickens; for she taketh sickness for sorrow of her chickens, and loseth her feathers. And her kindly love about her chickens is known by roughness of feathers, and by hoarseness of voice.

Bartholomew (Berthelet), bk. xii. § 18.

[N.B. In the article from which the above is an extract, the word "chickens" is spelt as follows: chekyns, chekens, chekynnes, chekennes, chykynnes, chykyns, and chykens.]

An odd number of eggs should always be put under a Hen, and that while the moon is waxing from the tenth to the fifteenth day. The flesh of hens clears the voice.

Hortus Sanitatis, bk. iii. ch. liii.

The Hens of country-houses have a certain ceremonious religion. When they have laid an egg, they fall a trembling and quaking, and all to shake themselves. They turn about also, as in procession, to be purified, and with some festue [or fescue, a straw] or such like thing, they keep a ceremony of hallowing, as well themselves as their eggs.

Holland's Pliny, bk. x. ch. xli.

If it thunder while she is broody the eggs will be addle, yea, and if the Hen chance but to hear an hawk cry they will be marred. The remedy against thunder is to put an iron nail under the straw of the Hen's nest, or else some earth newly turned up with the plough.

Ibid., ch. liv.

At this day, the English inhabitants eat almost no flesh more commonly than Hens.

Fynes Moryson, "Itinerary," bk. iii. ch. iii.

V. **Fowl.**

Herb.

Seeds and Herbs for the Kitchen.

Avens—Betony—Bleets or Beets, white or yellow—Bloodwort—Bugloss—Burnet—Borage—Cabbage, remove in June—Clary—Coleworts—Cresses—Endive—Fennel—French Mallows—French Saffron, set in August—Lang de Beef—

Leeks, remove in June—Lettuce, remove in May—Longwort—Liverwort—Marigolds, often cut—Mercury—Mints, at all times—Nep—Onions, from December to March—Orach or Arache, red and white—Patience—Parsley—Penny-royal—Primrose—Poret—Rosemary, in the spring-time, to grow south or west—Sage, red and white—English Saffron, set in August—Summer Savory—Sorrell—Spinach—Succory—Siethes—Tansey—Thyme—Violets of all sorts.

Herbs and Roots for Salads and Sauce.

Alexanders, at all times—Artichokes—Blessed Thistle, or Carduus Benedictus—Cucumbers in April and May—Cresses, sow with lettuce in spring—Endive—Mustard-seed, sow in the spring, and at Michaelmas—Musk-melon, in April and May—Mints—Purslane—Radish, and often remove them—Rampions—Rocket, in April—Sage—Sorrel—Spinach, for the summer—Sea-holly [*i.e.*, Eringo] Sparage [*i.e.*, Asparagus], let grow two years, and then remove—Skirrets, set these plants in March—Succory—Tarragon, set in slips, in March—Violets, of all colours.

These buy with the penny
Or look not for any.

Capers—Lemons—Olives—Oranges—Rice—Samphire.

Herbs and Roots, to boil or to butter.

Beans, set in winter—Cabbages, sow in March, and often remove—Carrots—Citrons, sow in May—Gourds, in May—Navews, sow in June—Pompions, in May—Parsnips, in winter—Runcival Pease, set in winter—Rapes, sow in June—Turnips in March and April.

Strewing Herbs of all Sorts.

Basil, fine and bushed, sow in May—Balm, set in March—Camomile—Costmary—Cowslips and Paggles [or Paigles, *i.e.*, Oxlips]—Daisies of all sorts—Sweet Fennel—Germander—Hyssop, set in February—Lavender—Lavender spike—Lavender cotton—Marjoram, knotted, sow or set, at the spring—Maudlein [*i.e.*, *Ageratum*, akin to Costmary, *i.e.*, *Balsamita*]—Penny-royal—Roses of all sorts, in January and September—Red Mints—Sage—Tansey—Violets—Winter Savory.

Herbs, Branches and Flowers, for Windows and Pots.

Bays, sow or plant in January—Bachelor's Buttons—Bottles, blue, red and tawny [*i.e.*, Corn-flowers]—Columbines—Campions—Cowslips—Daffodils, or Daffadowndillies—Eglantine, or Sweet-briar—Feverfew—Flower Amour, sow in May [*i.e.*, *Amaranthus*]—Flower de Luce—Flower Gentle, white and red [also *Amaranthus*]—Flower Nice—Gilliflowers, red, white, and carnations, set in spring, and at harvest in pots, pails, or tubs, or for summer in beds—Hollyhocks, red, white, and carnations—Indian Eye, sow in May, or set in slips in March—Lavender, of all sorts—Lark's Foot—*Laus Tibi*—*Lilium Convallium*—Lilies, red and white, sow or set in March or September—Marigolds double—*Nigella Romana*—Pansies, or Heart's-ease—Paggles, green and yellow—Pinks of all sorts—Queen's Gilliflowers—Rosemary—Roses of all sorts—Snap-dragon—Sops-in-wine [*i.e.*, Pinks]—Sweet Williams—Sweet Johns—Star of Bethlem—Star of Jerusalem—Stock Gilliflowers of all sorts—Tuft Gilliflowers—Velvet Flowers, or French Marigolds—Violets, yellow and white—Wall Gilliflowers of all sorts.

Herbs to still in Summer.

Blessed Thistle—Betony—Dill—Endive—Eyebright—Fennel—Fumitory—Hyssop—Mints—Plantain—Roses, red and damask—Respies [Raspberries]—Saxifrage—Strawberries—Sorrel—Succory—Woodruff, for sweet waters and cakes.

Necessary Herbs to grow in the Garden, for Physic, not rehearsed before.

Anise—Archangel—Betony—Chervil—Cingfoil—Cummin—Dragons—Dittany, or Garden Ginger—Gromwell seed, for the stone—Hart's Tongue—Horehound—Lovage, for the stone—Licquorice—Mandrake—Mugwort—Peony—Poppy—

Rue—Rhubarb—Smallage, for swellings—Saxifrage, for the stone—Savin, for the bots—Stitchwort—Valerian—Woodbine.

Thus ends in brief
Of Herbs the chief.
To get more skill
Read whom ye will
Such mo to have
Of field go crave.

Tusser's "Five Hundred Points of Good Husbandry," March's abstract.

I HAVE brought here good Herbs, and of them plenty,
To make good broth and farcing, and that full dainty.
. . . Here is Thyme and Parsley, Spinach and Rosemary,
Endive, Succory, Lacture, Violet, Clary,
Liverwort, Marigold, Sorrell, Hart's Tongue, and Sage,
Pennyroyal, Purslane, Bugloss and Borage,
With many very good Herbs, mo than I do name.

"The History of Jacob and Esau," iv. 5.

BE not merry among those that put Bugloss in their wine and sugar in thine.

Lilly, "Sappho and Phaon," ii. 1.

SUCH unexpected kindness
Is like Herb John in broth—
'T may e'en as well be laid aside as used.

"A Warning for Fair Women," Act i., line 331.

Herring.

TWELFTH NIGHT, iii. 1, 40.

THE Herring's eyes shine by night in the sea like a light, but their virtue dies with the fish. Wherever they see a light in the sea above the water, thither they swim in shoals. The Herring is said to live on water only, as the salamander on fire. The Herring helps against the bite of a dog, and of a sea-dragon. *Hortus Sanitatis*, bk. iv. § 3.

FRESH Herring plenty, Michell [*i.e.*, Michaelmas] brings
With fatted crones, and such old things.

Tusser, "Farmer's Daily Diet."

[The Dutch caught the Herrings in English waters, and sold them to Englishmen, so that they were sold in England at 20s. to 30s. the barrel; *cf.* "England's Way to Win Wealth" (1614); but "the English export into Italy great quantity of red Herrings" (*Fynes Moryson's* "Itinerary," bk. iii. p. 148), though this trade was afterwards encroached on by the Dutch.

As to the cooking of the red Herring: "Take well in worth a farthing-worth of flour to white him over and wamble him in" (*Nash's* "Lenten Stuff"). He was "hosted, roasted and toasted" (*ibid.*), "powdered and salted" (*ibid.*), and was served with mustard (*ibid.*, and *Greene's* "Looking-Glass for London," etc.), or with "oil and onion, crowned with a lemon-pill" (*Beaumont and Fletcher's* "Elder Brother"). The first dish that was brought up to table (at Queen's College, Oxford) on Easter Day was a red Herring riding away on horseback, *i.e.*, a Herring ordered by the cook something after the likeness of a man on horseback set in a corn salad (*Aubrey's* MS. Account of English Customs (1678).]

Hind. *V.* Deer.

Hog. *V.* Swine.

Honey.

KING HENRY V., i. 2, 199.

PHYSICIANS tell, that treat of kind of things, that Honey is unprofitable meat and grievous to children and young men, in the which is much heat, and according to full old men and cold, with wine and with hot meats. Also for Honey is even and temperate, Honey is much according and friend to kind, and likeneth itself much to the members. Honey keepeth and saveth and cleanseth and tempereth bitterness, and is therefore put in Conservatives, and cleanseth medicines to temper bitterness of spicery. But raw Honey not well clarified is right ventuous, and breedeth a fever that hight *Diurna*, and stretcheth and to-hauleth the body under the small ribs.

Bartholomew (*Berthelet*), bk. xix. § 54.

HONEY will suffer no dead bodies to putrefy.

Honey boiled cureth the wounds inflicted by the sting or teeth of serpents, and helpeth those who have eaten venomous mushrooms. Good it is also for to kill lice and such like vermin in the head, and to rid away nits.

Holland's Pliny, bk. xxii. ch. xxiv.

HONEY is engendered naturally in the air, and especially by the influence and rising of some stars. Be it what it will, either a certain sweat of the sky, or some unctuous jelly proceeding from the stars, or rather a liquor purged from the air when it purifieth itself; would God we had it so pure, so clear, and so natural, and in the own kind refined, as when it descendeth first, whether it be from sky, from star or from the air. *Ibid.*, bk. xi. ch. xii.

SOMETIME among honey deep in the hive, breedeth certain small worms, as it were attercops [spiders], and do spin and weave and make webs, and have the mastery of all the hive, and therefore the Honey rotteth and is corrupt. Honey that long abideth in old wax, waxeth red, and the corruption of Honey is like to the corruption of wine in flaskets [*i.e.*, bottles; *Bartholomew* has *in viribus*—in strength], and shall therefore be taken in time. Also bees do sit on the hive and suck the superfluity that is in the Honey-combs; and if they did not so the Honey should be corrupt that is in the combs, and spiders should be gendered. They sit on the combs, and do keep busily that those spiders have no mastery, and eat them if they find them, and should else all die.

Bartholomew (*Berthelet*), bk. xix. § 55.

OUR Honey is reputed and taken to be the best, because it is harder, better wrought and cleanlier vesselled up, than that which cometh from beyond the sea, where they stamp and strain their combs, bees and young blowings altogether into the stuff. Also it breedeth (being gotten in harvest-time) less choler. Our hives are made commonly of rye-straw, and wattled about with bramble quarters; but some make the same of wicker, and cast them over with clay. We cherish none in trees, but set our hives somewhere on the warmest side of the house. This furthermore is to be noted, that of Honey the best which is heaviest and moistest is always next the bottom.

Holinshed, "Description of England," p. 229.

V. Bee.

Honey-suckle.

MUCH ADO ABOUT NOTHING, iii. 1, 8.

FLIES that die on the Honey-suckle become poison to bees.

Lilly, "Sappho and Phaon," ii. 4.

Horse.

HORSES be joyful in fields, and smell battles, and be comforted with noise of trumps to battles and to fighting, and be excited to run with noise that they know, and be sorry when they be overcome, and glad when they have the mastery. And so feeleth and knoweth their enemies in battles, so far forth that they arise on their enemies with biting and smiting, and also some know their own lords, and forgetteth mildness, if their lords be overcome. And many Horses weep when their lords be dead. Also oft men that shall fight take evidence and divine and guess what shall befall by sorrow or by the joy that the Horse maketh. And those Horses be accounted best in war and in battle, that thrust the head deepest into the water when they drink. Also the gall of a Horse is accounted among venom. His fresh blood and raw is venomous, as the blood of a bull. The Horse's foam drunken with asses' milk slayeth venomous worms. Also sometime Horses have the podagre, and lose the soles of their feet, and then gendereth new. And sometime an Horse is wood [*i.e.*, mad], and the token thereof is that his ears bend toward the neck; and this evil hath no medicine. And the Horse knoweth his neighing that will fight with him, and hath liking to stand in meads, and to swim in water, and to drink troublous and thick water, and if the water be clear, the horse stampeth and stirreth it with his foot to make it thick. *Bartholomew* (*Berthelet*), bk. xviii. § 39.

I KNEW two [scholars] hired for ten groats apiece to say service on Sunday, and that's no more than a post Horse from hence [*i.e.*, from Rochester] to Canterbury [= 26 miles]. *Lilly*, "Mother Bombie," iv. 1.

[BUT in the time of the plague (1625) Horse-hire was dear:] "Coach-men ride a cock-horse and are so full of jadish tricks that you cannot be jolted six miles from London under 30 or 40 shillings."

Dekker, "A Rod for Run-aways (*Epistle to the Reader*).

[Horse-feed cost sixpence a day in Middleton's time ("Phœnix," i. 4, 35).]

[Ten pounds was a great price for a Horse (*cf.* *Dekker's* "Seven Deadly Sins"); and from the same author's "Bellman of London" we find that sales of Horses were registered in a toll-book. In the same tract is an account of "Horse-priggers" and "Horse-coursers," of whose tricks more is said in "Lanthorn and Candlelight," as well as of the cheating of hostlers, and of the sale of Horses at Smithfield.

In *Ruggles'* "Ignoramus" (First Prologue) a list of the favourite Race-Horses in 1614-15 is given.

Sir Thos. Browne devotes a whole chapter of his "Vulgar Errors" to a refutation of the fallacies that Horses have no gall, and that, if they have gall, it is venomous.]

IT is said that if Horses be shod with that iron, wherewith one hath been before killed, it makes the same Horses very lively and quick. And if of the same you make a bit or a snaffle, that Horse that hath it in his mouth will be made tame and easy to be handled, yea, though he be never so wild, stubborn, or given to biting.

Lupton, "A Thousand Notable Things," bk. vii. § 97.

IN the heart of Horses there is found a bone most like unto a dog's tooth; it is said that this doth drive away all grief or sorrow from a man's heart, and that a tooth being pulled from the cheeks or jaw-bones of a dead

Horse doth shew the full and right number of the sorrows of the party so grieved. If swords, knives, or the points of spears when they are red fire hot, be anointed with the sweat of a horse, they will be so venomous and full of poison, that if a man or woman be smitten or pricked therewith, they will never cease from bleeding as long as life doth last. *Topsell*, "Four-footed Beasts," pp. 337-8.

THE tooth of a yearling colt laid on the neck of a baby, makes its teeth come without pain.

Albertus Magnus, "Of the Wonders of the World."

THE tooth of a mare placed on the head of a raving madman straightway frees him. *Ibid.*

THE hoof of a Horse burnt in a house drives away mice. The same with the hoof of a mule. *Ibid.*

THE Londoners pronounce woe to him that buys a Horse in Smithfield, that takes a servant in Paul's Church, that marries a wife out of Westminster.

Fynes Moryson, "Itinerary," part iii. p. 53.
Cf. ii. KING HENRY IV., i. 2.

OUR Horses moreover are high, and although not commonly of such huge greatness as in other places of the main, yet if you respect the easiness of their pace, it is hard to say where their like are to be had. Such as serve for the saddle are now grown to be very dear among us. There is no greater deceit used anywhere than among our Horse-keepers, Horse-coursers and ostlers. There are certain notable markets, wherein great plenty of Horses and colts is bought and sold, as Ripon, Newport Pond, Wolfpit, Harborough and divers other. But as most drovers are very diligent to bring store of these unto those places, so many of them are too lewd in abusing such as buy them. For they have a custom to make them look fair to the eye, — when they come within two days' journey of the market, to drive them till they sweat, and for the space of 8 or 12 hours, which being done, they turn them all over the backs into some water, where they stand for a season, and then go forward with them to the place appointed, where they make sale of their infected ware, and such as by this means do fall into many diseases and maladies. Of such outlandish Horses as are daily brought over unto us I speak not, as the jennet of Spain, the courser of Naples, the hobby of Ireland, the Flemish roil and Scottish nag. King Henry VIII. erected a noble studdery, and for a time had very good success with them, till the officers waxing weary procured a mixed brood of bastard races, whereby his good purpose came to little effect. Sir Nicholas Arnold of late hath bred the best horses in England.

Holinshed, "Description of England," p. 220.

Horse-leech.

KING HENRY V., ii. 3, 58.

IN a river of Mauritania are found some of seven cubits in length, which breathe, through perforations in the gullet. Leeches are produced from rottenness, and it is not known whether they gender.

Jonston's "Natural History," bk. iv. ("On Insects"), tit. ii. ch. i.

Hound. *V.* Dog.

Hyæna.

AS YOU LIKE IT, iv. 1, 156.

HYÆNA is a cruel beast like to the wolf in devouring and gluttony. It is his kind to change sexes, for he is now found male and now female, and is therefore an unclean beast. And cometh to houses by night, and feigneth man's voice as he may, for men should trow that it is a man. And herds tell that among stables, he feigneth speech of mankind, and calleth some man by his own name, and rendeth him when he hath him without. And he feigneth oft the name of some man for to make hounds run out, that he may take and eat them. And hath the neck of the adder viper, and the ridge of an elephant, and may not bend but if he bear all the body about. And this beast hath endless many manners and diverse colours in his eyes, and full movable eyes and unsteadfast. And his shadow maketh hounds leave barking, and be still, if he come near them. And if this beast Hyæna goeth thrice about any beast, that beast shall stint [*i.e.*, stop] within his steps. And this beast gendereth with a lioness of Ethiopia, and gendereth on her a beast that is most cruel, and followeth the voice of men and of tame beasts, and hath many rows of teeth in every side of the mouth. This beast Hyæna breedeth a stone that hight Hyæna; and what man that beareth it under his tongue, he shall by virtue of that stone divine and tell what shall befall. Also Hyæna hateth the panther. And if both their skins be hanged together, the hair of the panther's skin shall fall away. This beast Hyæna fleëth the hunter, and draweth toward the right side to occupy the trace of the man that goeth before; and if he [*i.e.*, the man] cometh not after, he [the man again] goeth out of his wit, or else falleth down off his horse. And if he turn against the Hyæna, the beast is soon taken. And also witches use the heart of this beast and the liver in many witchcrafts.

Bartholomew (*Berthelet*), bk. xviii. § 61.

In the Hyæna itself there is a certain magical virtue transporting the mind of man or woman, and ravishing their senses so as that it will allure them unto her very strangely. When the Hyænas fly before the hunter and would not be taken, they wind with a career out of the way toward the right hand, and wheel about until the man be gotten before them; and this they do because they would meet with his tracts and footing; which if they happen upon, and get behind him, you shall see the hunter incontinently to be so intoxicate in his brain, that he is not able to bear his head nor sit his horse, but to fall from his back. But in case that they turn on the left hand, it is an evident sign that they be ready to faint, and then will they quickly be taken. The sooner also and with more ease be they caught if the hunter tie his girdle about his middle with 7 knots, and the cord of his whip likewise wherewith he ruleth and jerketh his horse with as many. This chase after the Hyæna must be just at the very point when the moon is passing through the sign Gemini ; and then if they be taken, the huntsman must be sure to save every hair of their skins, and miss not one, so medicinable they are. Whosoever are haunted with sprites in the night-season, and be affrighted with such bugbears, let them but take one of the master-teeth of the Hyæna, and wear it about them tied by a linen thread, they shall be freed from all such fantastical illusions. And as for those that wear under the soles of their feet within the shoe a Hyæna's tongue, there is not a dog will be so hardy as to bay or bark at them. And the hairs growing about the muzzle of this beast have an amatorious virtue with them to make a woman love a man, in case her lips be but touched therewith. If the side-posts or door-cheeks of any house be striked with the Hyæna's blood, wheresoever magicians are busy with their feats and juggling casts, they shall take no effect, whether they be charms, exorcisms or invocations; insomuch as they shall not be able to raise up spirits, nor have any conference with familiars by any means of conjuration, whether it be by torch-lights, by bason, by water, by globe or otherwise. A decoction made with the ashes of the pastern bone of the left leg, boiled together with the blood of a weasel, causeth as many as be anointed therewith to be odious in the eyes of all men. The hindmost end of the gut in this beast is of virtue that no captain, prince or potentate shall be able to wrong or oppress those who have but the same about them ; but contrariwise assureth them of good speed in all their petitions, and of happy issue in all suits of law and trials of judgments. *Holland's Pliny*, bk. xxviii. ch. viii.

THE Hyæna when she mourns is then most guileful.

"Euphues' Golden Legacy."

THE middle of his back is a little crooked or dented, the colour yellowish, but bespeckled on the sides with blue spots, which make him look more terrible, as if it had so many eyes. The eyes change their colour at the pleasure of the beast, a thousand times a day. The skilful lapidarists affirm that the beast hath a stone in his eyes (or rather in his head) called Hyæna or Hyænius ; but the ancients say, that the apple or pupil of his eye is turned into such a stone, and that if a man lay it under his tongue, he shall be able to foretell and prophesy of things to come. Their neck cannot bend, except the whole body be turned about. This beast hath a very great heart. There is a fish of this name which turneth sex. [Hyænas] engender not only among themselves, but also with dogs, lions, tigers and wolves. This is accounted a most subtle and crafty beast, and the female is far more subtle than the male, and therefore more seldom taken, for they are afraid of their own company. If she find a man or dog on sleep she [kills it if it be smaller than herself, and runs away if it be bigger]. One of these coming to a man asleep in a sheep-cote, by laying her left hand or fore-foot to his mouth, made or cast him into a dead sleep, and afterward digged about him such a hole like a grave, as she covered all his body over with earth, except his throat and head, whereupon she sat, until she suffocated and stifled him; yet this is attributed to her right foot. There is also great hatred between a pardel and this beast, for if after death their skins be mingled together, the hair falleth off from the pardel's skin, but not from the Hyæna's. He that will go safely through the mountains or places of this beast's abode must carry in his hand a root of coloquintida. Also if a man compass his ground about with the skin of a crocodile, an Hyæna, or a sea-calf, and hang it up in the gates or gaps thereof, the fruits enclosed shall not be molested with hail or lightning. And a man clothed with this skin may pass without fear or danger through the midst of his enemies. A fig-tree also is never oppressed with hail or lightning; and the true cause hereof is assigned by the philosophers to be the bitterness of it; for the influence of the heavens hath no destructive operation upon bitter, but upon sweet things. If the left foot and nails be bound up together in a linen bag, and so fastened unto the right arm of a man, he shall never forget whatsoever he hath heard or knoweth. And if he cut off the right foot with the left hand and wear the same, whosoever seeth him shall fall in love with him, besides the beast. Also the marrow of the right foot is profitable for a woman that loveth not her husband, if it be put into her nostrils. And with the powder of the left claw, they which are anointed therewith, it being first of all decocted in the blood of a weasel, do fall into the hatred of all men. And if the nails of any beast be found in his maw after he is slain, it signifieth the death of some of his hunters. The dung or filth of an Hyæna, being mingled with certain other medicines is very excellent to cure and heal the bites and stingings of crocodiles, and other venomous serpents.

Topsell, "Four-footed Beasts," pp. 339-47.

Hyssop.

OTHELLO, i. 3, 325.

IN summer when Hyssop beareth flowers, ye must gather them, and dry them in a clean place and dark, that it be not smoky, and they have virtue to dissolve, to temper, to consume, to waste and to cleanse the lungs.

Bartholomew (*Berthelet*), bk. xvii. § 85.

HYSSOP, stamped with honey, salt and cummin, and so reduced into a plaster, is thought to be a proper remedy for the sting of serpents. *Holland's Pliny*, bk. xxv. ch. 11.

IF a man perceive that he hath either inwardly taken for a medicine, or applied outwardly, a radish root which is over strong, he must presently have Hyssop given him; for this antipathy and natural contrariety there is between these two herbs, that the one correcteth the other.

Ibid., bk. xx. ch. 4.

You are, Sir,
Just like the Indian Hyssop, prais'd of strangers
For the sweet scent, but hated of the inhabitants
For the injurious quality.

Robert Davenport, "City Night-cap," Act. i. (1624).

Incense or Frankincense.

KING LEAR, v. 3, 21.

[FRANKINCENSE] is the name of a tree, and of the gum that oozeth and cometh out thereof. It is a tree of Arabia, and is great with many boughs, and with the most lightest rind. And thereof cometh juice with good smell, and is white as almonds, and is fat when it is tempered and neshed [*i.e.*, softened]. And so the tree that beareth Frankincense groweth without tilling, and loveth clay-land; and the Arabs tell that Frankincense shall not be gathered nor the tree thereof pared but of holy men and religious, that be not defiled by touching of women in time of gathering. Frankincense is gathered and brought on camels' backs to the city that hight Sabocriam; and there is a gate opened therefor. And it is not lawful to lead it by another way. And [it] is not lawful to beg neither to sell thereof, before due portion be offered to the god that they worship. And is assayed by witness, if it burneth anon to coals, and waxeth on light on high, if it hold not together the teeth, when it is bitten, but breaketh anon and falleth to powder. Of Frankincense set afire cometh a good smelling smoke, shapen as a rod, and small beneath, and full movable, and turning, and crooked with many bendings and wrinklings, and moveth towards contrary sides with most light movings, and destroyeth stench of carrion by good savour thereof, and thirleth and passeth straight to the brain, and comforteth and refresheth the spirit of feeling, and spreadeth into the cells of the brain.

Bartholomew (*Berthelet*), bk. xvii. § 173.

If Incense be drunk by a healthy man, he runs the risk of becoming mad, or of dying. It strengthens the memory.

Hortus Sanitatis, bk. i. ch. cccclxxxiv.

In English Frankincense and Incense. It doth help and strengthen the wit and understanding, but the often taking of it will breed the head-ache, and if too much of it be drunk with wine, it killeth.

Gerard's "Herbal," *s.v.*

Insane Root.

Macbeth, i. 3, 85.

[The commentators consider this to be hemlock, but it is possible that it might be henbane, which as Gerard notes ("Herbal," *s.v.*), was called *Insana*, and, according to Pliny (bk. xxv. ch. 4), troubles the brain, and puts men beside their right wits.]

Ivy.

MIDSUMMER NIGHT'S DREAM, iv. 1, 48.

IVY multiplieth milk in goats that eat thereof. The root thereof pierceth things that be full hard; and is cold of kind, and tokeneth that the ground is of cold kind that it groweth in. And of Ivy is double kind, white and black, male and female; the male is harder in leaves and more fat and greater. The white Ivy hath white fruit, and the black hath black. The shadow thereof is noyful and grievous, and strong enemy to cold, and most loved of serpents, and breaketh walls and graves. Also the kind of Ivy is full wonderful in knowledge and assaying of wine. For it is certain, that if wine meddled with water be in a vessel of Ivy, the wine fleeth over the brink, and the water abideth. *Bartholomew* (*Berthelet*), bk. xvii. § 53.

THE gum of Ivy killeth lice and nits, and being laid to it taketh away hair. It is unwholesome to sleep under the Ivy or in an Ivy-bush. It maketh the head light and dizzy. *Batman* on *Bartholomew, ut supra.*

ALTHOUGH Ivy be cut asunder in many places, yet it continueth and liveth still.

Holland's Pliny, bk. xvi. ch. 34.

THE liquor issuing out of Ivy is depilatory; but as it taketh away hair, so it riddeth lice and vermin. The berries of Ivy colour the hair black. The juice of the Ivy-root, drawn with vinegar and taken in drink, is singular against the poison of the venomous spiders Phalangia.

Ibid., bk. xxiv. ch. 10.

BOARS cure their ailments with Ivy. A man crowned with Ivy cannot get drunk.

Hortus Sanitatis, bk. i. § 172.

CATO saith that a cup of Ivy will hold no wine at all. I have made some vessels of the same wood, which refuse no kind of liquor; and yet I deny not but the Ivy of Greece or Italy may have such a property.

Holinshed, "Description of England," p. 239.

Jack-a-napes.

KING HENRY V., v. 2, 148.

YOUR wife is your ape, and that heavy burden, wedlock, your Jack an ape's clog.

Dekker, "Patient Grissell," line 814.

HE would sit upon's tail before [my enemies], and frown like John-a-Napes when the Pope is named.

Thomas Killigrew, "The Parson's Wedding," v. 2.

V. Ape.

Jet.

MERCHANT OF VENICE, iii. 1, 42.

JET is a boisterous [*Bartholmew—rudis*] stone, and nevertheless it is precious. Most plenty and best be in Britain. And is double, that is to say, yellow and black. The black is plain and light, and burneth soon in fire, and driveth away adders with smell thereof, when it is kindled. This giveth monition of them that have fiends within them; and is holden contrary to fiends; and giveth knowledge of maidenhead,—for if a maid drink of the water thereof, *non urinabit*, and if she be no maid and drinketh thereof, *urinabit* anon and also against her will; and so by this stone a maiden is anon proved. Also the power thereof is good to feeble teeth and wagging, and strengtheth and fasteneth them. Also it is said that this stone helpeth for fantasies, and against vexations of fiends by night; also it helpeth against witch-craft, and fordoeth [hinders] hard enchantments. And so, if so boisterous a stone doth so great wonders, none should be despised for foul colour without, while the virtue that is hid within is unknown. And this stone is kindled in water, and quenched in oil, and that is wonder.

Bartholomew (*Berthelet*), bk. xvi. § 49.

EVEN to this day there is some plenty to be had of this commodity in Derbyshire and about Berwick, whereof rings, salts, small cups and sundry trifling toys are made. The German writers confound it with amber as if it were a kind thereof [because of its electrical property]. Charles IV.

Emperor glazed the church withal that standeth at the Fall of Tangra, but I cannot imagine that light should enter thereby. The writers also divide this stone into five kinds, of which the one is in colour like unto lion-tawny, another streaked with white veins, the third with yellow lines, the fourth is garled [variegated] with divers colours, among which some are like drops of blood (but those come out of Ind), and the fifth shining-black as any raven's feather.

Holinshed, "Description of England," p. 239.

Jewel.

MERCHANT OF VENICE, iii. 1, 91.

IN the earth are many kinds of humours. Some there are which are hardened by nature; from these all ores of metal are generated, from which gold and silver are made. Some there are which are turned from fluid into stone, from whence precious stones grow.

Hortus Sanitatis, bk. v. (Introduction).

WERE there no sun, by whose kind, lovely heat,
The earth brings forth those stones we hold of price.

Heywood, First Part of "King Edward IV."

[This same theory is alluded to in *Glapthorne's* "Hollander."]

HERE'S a Jewel for thee,
A pretty wanton label for thine ear.

Dekker, "Witch of Edmonton," iii. 2.

I WOULD have a Jewel for mine ear,
And a fine brooch to put into my hat.

Marlowe, "Dido, Queen of Carthage," i. 1.

A DIAMOND ring out of her ear.

Randolph, "Jealous Lovers," i. 8 (stage direction).

[So also *Beaumont and Fletcher's* "A King and No King," i. 1, and *Ben Jonson's* "Every Man in his Humour," iv. 9, etc. In the same author "Fox," (ii. 5) we find:

WERE you enamoured on his copper rings
His saffron Jewel, with the Toad-stone in't?

V. Gem.

Keech.

ii. KING HENRY IV., ii. 1, 101.

A KEECH is the fat of an ox rolled up by the butcher into a round lump.

Steevens, loc. cit., and *cf.* KING HENRY VIII. i. 1, 55.

Kernel. *V.* Nut.

Kex.

KING HENRY V., v. 2.

V. **Hemlock.**

Kite.

A KITE is weak in flight and in strength. And is a bird that may well away with travail, and therefore he taketh cuckoos upon his shoulders, and beareth them, lest they fail in space of long ways, and bringeth them out of the countries of Spain. And he is a ravishing fowl, and hardy among small birds, and a coward and fearful among great birds, and dreadeth to lie in wait to take wild birds, and dreadeth not to lie in wait to take tame birds, and lieth oft in wait to take chickens, and he eateth carrions and unclean things. And is taken with the sparhawk, and for his faintness and cowardness he is overcome of a bird that is less than he. And in youth there seemeth no difference between the Kite and other birds of prey; but the longer he liveth, the more he sheweth that his own kind is unkind. And there is a manner Kite, that taketh birds in the beginning, and afterward he eateth guts of beasts, and taketh unneath afterward flies and small worms. And he dieth for hunger at the last, and is a cruel fowl about his birds [*i.e.*, young], and is sorry when he seeth them fat; and to make them lean, he beateth them with his bill, and withdraweth their meat; and hath a voice of plaining and of moan, as it were messenger of hunger; for when he hungereth, he seeketh his meat weeping with voice of plaining and of moan.

Bartholomew (Berthelet), bk. xii. § 26.

THE Kites or Gleads are of the kind of hawks or birds of prey, only they be greater. These Gleads or Puttocks seem by the winding and turning of their tails to and fro as they fly to have taught pilots the use of the helm. After the sunsteads alway in summer they be troubled with the gout in their feet. *Holland's Pliny*, bk. x. ch. 10.

IF the Kite's head be taken, and borne before the breast, it brings the love and favour of all men and women; if it be hung on a hen's neck, it will never cease to run, until it gets it off; and if a cock's comb be anointed with its blood, from thenceforth it will not crow. In its knees is found a certain stone, if you look carefully, and if this be put in the food of two enemies, they will become friends, and there will be good peace between them.

Albertus Magnus, "Of the Virtues of Animals."

Knot-grass [*Polygonum Vulgare*].

Hindering knot-grass.
MIDSUMMER NIGHT'S DREAM, iii. 2, 239.

WE want a boy extremely for this function kept under for a year with milk and Knot-grass.

Beaumont and Fletcher, "The Coxcomb," ii. 1.

[Steevens quotes also to the same effect, "The Knight of the Burning Pestle," by the same authors, but I have not the reference.

Knot-grass is binding, stays any flux, solders the lips of green wounds, and knits broken bones, according to *Parkinson's* "Herbal."]

Lady-smock (or Cuckoo-flower).

LOVE'S LABOUR'S LOST, v. 2, 905.

V. Cuckoo-bud.

Lamb.

AMONG all the beasts of the earth, the Lamb is most innocent, soft and mild; for he nothing grieveth neither hurteth nor with teeth nor with horn nor with claws. Those which be yeaned in springing time be more huge and great of body, and more stronger of body, than those which be yeaned in harvest and in winter. Lambs which be conceived in the Northern wind be better than those that be conceived in the Southern wind. And Lambs have such colour in flesh and in wool, as the father and the mother have colour in veins of the tongue. The Lamb hoppeth and leapeth tofore the flock, and playeth, and dreadeth full sore when he seeth the wolf, and fleeth suddenly away; — but anon he is astonied for dread, and stinteth [*i.e.*, stoppeth] suddenly, and dare flee no further; and prayeth to be spared, not with bleating, but with a simple cheer, when he is taken of his enemy. Also whether he be led to pasture or to death, he grudgeth not, nor pranceth not, but is obedient and meek. It is peril to leave Lambs alone, for they die soon, if there fall any strong thunder; for the Lamb hath kindly a feeble head.

Bartholomew (*Berthelet*), bk. xviii. § 4.

LAMBS have an evil, that is when they be too fat about the reins, for if the tallow covereth the reins, then they die, and the tallow increaseth in good pasture; and, therefore, Lambs be put out of the pasture, lest they wax too fat.

Ibid., § 6.

THE rennet of a Lamb is good against all evil medicines, and against the bites or blows of marine animals, and cures all venomous bites.

Hortus Sanitatis, bk. ii. § 2.

Lapwing.

MEASURE FOR MEASURE, i. 4, 32.

THE Lapwing eateth man's dirt; for it is a bird most filthy and unclean, and is copped [*i.e.*, crested] on the head, and dwelleth always in graves or in dirt. And if a man anoint himself with her blood when he goeth to sleep, he shall see fiends busy to strangle and snare him; and her heart is good to evil-doers, for in their evil-doings they use their hearts. When he ageth, so that he may neither see nor fly, his birds [*i.e.*, young] pull away the feeble feathers, and anoint his eyes with juice of herbs, and hide him under their wings till his feathers be grown; and so he is renewed, and flieth and seeth clearly.

Bartholomew (*Berthelet*), bk. xii. § 37.

THE Lapwing's tongue, if hung over a man who suffers from much forgetfulness, helps him.

Hortus Sanitatis, bk. iii. § 118.

IF the Lapwing do sing before the vines do bud, it foreshows great plenty of wine.

Lupton, "A Thousand Notable Things," bk. ix. § 21.

QUIRIN is a stone that is found in Lapwings' nests. This stone bewrayeth and discovereth in sleep counsel and privity. For this stone, laid and set under a man's head that sleepeth, maketh him tell as he thinketh sleeping, and multiplieth wonderly phantasies. Therefore witches love that stone, for they work witchcraft therewith.

Bartholomew (*Berthelet*), bk. xvi. § 83.

Latten.

MERRY WIVES OF WINDSOR, i. 1, 165.

LATTEN, though it be brass or copper, yet it shineth as gold without. Also Latten is hard as brass or copper. For by meddling of copper and of tin and of *auripigmentum*, and with other metal, it is brought in the fire to colour of gold. Also of Latten be composed divers manner of vessel, and fair, that seem gold when they be new, but the first brightness dimmeth some and some, and becometh as it were rusty. *Bartholomew* (*Berthelet*), bk. xvi. § 5.

Laurel.

iii. KING HENRY VI., iv. 6, 34.

THE common fame is that only this tree is not smit with lightning; therefore the land that beareth laurel-tree is safe from lightning both in field and in house. The green leaves thereof that smell full well, if they be stamped, healeth stinging of bees and of wasps.

Bartholomew (*Berthelet*), bk. xvii. § 48.

V. Bay.

Lavender.

WINTER'S TALE, iv. 3, 104.

LAVENDER has no seed. The smell thereof oppresses the head, and causes sleep. *Hortus Sanitatis*, bk. i. § 250.

THOSE hose are in Lavender [*i.e.*, pawned].

Middleton, "Family of Love," iii. 2, 79.

[So *Ben Jonson*, "Every Man out of his Humour," iii. 3, *Greene*, "Quip for an Upstart Courtier."]

Lead.

OF brimstone that is boistous, and not swiftly pured, but troubly and thick, and of quicksilver, the substance of lead is gendered; so of uncleanness of unpure brimstone lead hath a manner neshness [softness], and smircheth his hand that toucheth it. If thou hang lead over vinegar, it hurteth it; for vinegar shall thirl the substance thereof, and turn it to powder; therewith women paint themself for to seem fair of colour.

Bartholomew (*Berthelet*), bk. xvi. § 81.

TIN and Lead are very plentiful with us, the one in Cornwall, Devonshire and elsewhere in the North, the other in Derbyshire, Weardale, and sundry places of this island, whereby my countrymen do reap no small commodity, but especially our pewterers. There were mines of lead sometimes also in Wales, which endured so long till the people had consumed all their wood by melting of the same.

Holinshed, "Description of England," p. 238.

Leather-coat.

ii. KING HENRY IV., v. 3, 44.

[Henley, in Steevens' edition (*loc. cit.*), says: "The apple commonly denominated "russeting," in Devonshire is called the 'buff-coat.'"

Evelyn, in his "Kalendarium Hortense," under "Fruits in Prime, or yet lasting" (in December), enumerates, "Russeting, Pippins, Leather-coat"; and in the same work "Leather-coat" is distinguished from "russet pippin" and "golden russet pippin" in the catalogue of the best apple-trees.]

Leech. *V.* Horse-leech.

A WATER-LEECH sitteth upon venomous things, and, therefore, when he shall be set to a member bycause of medicine, first he shall be wrapped in nettles and in salt, and is thereby compelled to cast out of his body if he hath tasted any venomous thing in warm water.

Bartholomew (*Berthelet*), bk. xviii. § 93.

Leek.

The [leek] skin is good for your broken coxcomb.
KING HENRY V., v. 1, 55.

THE juice thereof drunk with wine helpeth against biting of serpents, and against every venomous beast. Leek stamped with honey healeth wounds, if it be laid thereto in a plaster-wise. Leek meddled with salt closeth soon and healeth new wounds, and soldereth soon breaches. And Leeks eaten raw helpeth against drunkenness. Also the smell of Leeks driveth away scorpions and serpents, and healeth the biting of a wood hound with honey, and breedeth sleep.

Bartholomew (*Berthelet*), bk. xvii. § 133.

IF you prick the head of a Leek with a reed or a stick sharped, and put within the same the seeds of rape, or of cucumbers, the said Leek's head will so swell, that it will seem monstrous.

Lupton, "A Thousand Notable Things," bk. ii. § 51.

Now Leeks are in season, for pottage full good,
And spareth the milch-cow, and purgeth the blood;
These having with peason, for pottage in Lent,
Thou sparest both oatmeal, and bread to be spent.

Tusser, "Good Husbandry," March, st. 26.

WHEN the seed of a Leek is thrown upon vinegar, it restores its acidity.

Albertus Magnus, "Of the Wonders of the World."

Lemon.

LOVE'S LABOUR'S LOST, V. 2, 653.

[Probably unknown to Bartholomew, who does not mention lemons nor oranges.]

IT is eaten seasoned with salt.

Hortus Sanitatis, bk. i. § 260.

THY breath smells of Lemon-pills.

Webster, "Duchess of Malfi," ii. 1.

IF you want Lemon-waters
Or anything to take the edge of the sea off,
Pray speak and be provided.

Beaumont and Fletcher, "Tamer Tamed," iv. 4.

V. also **Cloves**.

Leopard.

KING RICHARD II., i. 1, 174.

THE Leopard is a beast most cruel, and is gendered in spouse breach of a pard and of a lioness. The Leopard is a full resing [*i.e.*, raging] beast and headstrong, and thirsteth blood, and the female is more cruel than the male, and pursueth his prey startling and leaping and not running, and if he taketh not his prey in the third leap, or in the fourth, then he stinteth [*i.e.*, ceaseth] for indignation, and goeth backward, as though he were overcome. And he is less in body than the lion, and therefore he dreadeth the lion, and maketh a cave under earth with double entering, one by which he goeth in, and another by which he goeth out. And that cave is full wide and large in either entering, and more narrow and strait in the middle. And so when the lion cometh, he fleeth and falleth suddenly into the cave, and the lion pursueth him with a great rese [*i.e.*, rush, or rage], and entereth also into the cave, and weeneth there to have the mastery of the Leopard, but for greatness of his body he may not pass freely by the middle of the den, which is full strait,—and when the Leopard knoweth that the lion is so let and holden in the strait place, he goeth out of the den forward, and cometh again into the den in the other side behind the lion, and reseth on him behindforth with biting and with claws. And so the Leopard hath often in that wise the mastery of the lion by craft and not by strength. This beast eateth sometime venomous thing, and seeketh then man's dirt and eateth it, and therefore hunters hangeth such dirt in some vessel on a tree, and when the Leopard cometh to that tree, and leapeth up to take the

dirt, then the hunters slay him in the meantime, while he is thereabout. Also sometime the Leopard is sick, and drinketh wild goat's blood, and scapeth by it the sickness in that wise. *Bartholomew (Berthelet)*, bk. xviii. § 67.

THE Leopard flees when he sees a man's scull, and he is afraid of the grass called Leopard-grass, and is killed by the herb which is called "strangle-leopard" [perhaps aconite, —"Leopard's bane"]. *Hortus Sanitatis*, bk. ii. § 81.

Lettuce.

OTHELLO, i. 3, 324.

WHEN it is old, it is hard, and use thereof appaireth [injures] the sight, and maketh it fail, and slayeth the feeling, for it stifleth natural feeling with sourness thereof. A manner kind of Lettuce groweth of itself without tilling, and if it be thrown into the sea, it slayeth all the fish that is nigh thereabout. Hawks scrape this herb, and take out the juice thereof, and touch and heal their eyes therewith, and do away dimness and blindness when they be old. And it healeth biting of serpents, and stinging of scorpions, if the juice thereof be drunk in wine, and the leaves, stamped and laid to the wound in a plaster wise, suageth and healeth all manner swelling. But oft use thereof, and too much thereof eaten grieveth the clearness of the eyes.

Bartholomew (*Berthelet*), bk. xvii. § 92.

BY manuring, transplanting, and having a regard to the moon and other circumstances, the leaves of the artificial Lettuce are oftentimes transformed into another shape. If Lettuce be boiled, it is sooner digested, and nourisheth more. It is served in these days, and in these countries, in the beginning of supper, and eaten first before any other meat; for being taken before meat it doth many times stir up appetite; eaten after supper it keepeth away drunkenness which cometh by the wine.

Gerard's "Herbal," *s.v.*

WHEN I nursed thee with Lettuce, would it had turned to hemlock.

Lilly, "Sappho and Phaon," iv. 2.

Leviathan.

MIDSUMMER NIGHT'S DREAM, ii. 1, 174.

THE Leviathan often lies in wait for the whale, and fights with him; and all the fishes of the sea which behold the fight flock quickly to the tail of the whale. Now if the whale be overcome he must die, and those fish too, which he had girdled with his tail, are quickly swallowed. But if the Leviathan cannot overcome the whale, he emits from his jaws a most foul stench with water; but the

whale swallows the water, and rejects it, and repels that very foul stench, and so saves and defends him and his.

Hortus Sanitatis, bk. ii. § 84, and bk. iv. § 50.

Libbard.

LOVE'S LABOUR'S LOST, v. 2, 551.

V. Leopard.

Lily.

THE root thereof drunk with wine healeth biting of serpents, and helpeth against the malice and venom of frogs.

Bartholomew (*Berthelet*), bk. xvii. § 91.

IF the root be curiously opened, and therein be put some red, blue or yellow colour that hath no caustic or burning quality, it will cause the flower to be of the same colour.

Gerard's "Herbal," *s.v.*

THE leaves of the herb applied are good against the stinging of serpents. The roots boiled in wine causeth the corns of the feet to fall away within few days, with removing the medicine until it have wrought his effect.

Ibid.

It is called in English Lily of the Valley, or the Conval Lily, and May-Lilies, and in some places Liticonfancy. The flowers of the Valley Lily distilled with wine, and drunk, the quantity of a spoonful, restoreth speech unto those that have the dumb palsy, and that are fallen into the Apoplexy. The water aforesaid doth strengthen the memory that is weakened and diminished. The flowers of May-Lilies put into a glass, and set in a hill of ants close stopped for the space of a month and then taken out, therein you shall find a liquor that appeaseth the pain and grief of the gout, being outwardly applied, which is commended to be most excellent. *Ibid., s.v. Lilly in the Valley.*

If you gather this herb while the sun is in the Sign of Leo, and mix it with the juice of laurel, then put it under dung for some time, worms will be generated, and if a powder be made of these, and be strewed about the neck of anyone, or in his clothes, he will never sleep, nor be able to sleep, until it has been removed. And if you shall anoint anyone with these worms, he will straightway become feverish. And if the said plant be placed in any vessel in which there is cow's milk, and covered with the skin of a cow of one colour, all the cows will lose their milk. And this has been well tried in our time.

Albertus Magnus, "Of Virtues of Herbs," § 9.

Lime. *V.* Bird-lime.

Tempest, iv. 1, 246.

Lime.

Midsummer Night's Dream, v. 1, 132.
ii. King Henry IV., ii. 4, 136, etc.

Lime is called hot, for while it is cold in handling, it containeth privily within fire and great heat; and when it is sprung [*i.e.*, sprinkled] with water, anon the fire that is within breaketh out. In the kind thereof is some wonder; —for after that it is burnt, it is kindled in water that quencheth fire. Hot lime sod with auripigment and water maketh hair to fall. *Bartholomew* (*Berthelet*), bk. xvi. § 24.

SINCE the Spanish sacks have been common in our taverns, which for conservation are mingled with Lime in the making.

Sir Richard Hawkins, quoted in *Steevens'* Shakespeare.

Ling.

ALL'S WELL THAT ENDS WELL, iii. 2, 14.

[OF salted and dried cod-fish] Haberdine and Ling are accounted the best and daintiest.

Hart, "Diet of the Diseased," bk. i. ch. xxi.

RED herring and Ling never come to the board without mustard their waiting-maid. *Nash*, "Lenten Stuff."

Dank Ling forgot
Will quickly rot.

Tusser, "Good Husbandry," December's Abstract.

[A side of Ling cost 20d. in 1573; whereas a side of haberdine cost but 8d. A whole Ling was worth 3s. to 3s. 4d.]

HIS Majesty's [James I.'s] Serjeant-caterer hath yearly gratis, out of every ship and bark, one hundred of the choicest and fairest Lings, which are worth more than £10 the hundred.

Tobias Gentleman, "England's Way to Win Wealth."

A GOOD tough gentleman! He looks like a dry Poul [perhaps Jowl] of Ling upon Easter-Eve, that has furnished the table all Lent.

Ben Jonson, "Every Man out of his Humour," iv. 4.

Lion.

SOME Lions be short with crisp hair and mane, and these Lions fight not; and some Lions have simple hair of mane, and these Lions have sharp and fierce hearts. And he dreadeth noise and rushing of wheels, but he dreadeth fire much more. And when they sleep their eyes wake. And when they go forth or about, they hele [*i.e.*, cover or conceal] their fores [*i.e.*, goings] and steps, for hunters should not find them. And it is trowed that the Lion-whelp, when he is whelped, sleepeth three days and three nights; and the place of the couch trembleth and shaketh by roaring of the father, that waketh the whelp that sleepeth. It is the kind of Lions not to be wroth with man, but if they be grieved or hurt. Also their mercy is known by many and oft ensamples; for they spare them that lie on the ground, and suffer them to pass homeward that be prisoners, and come out of thraldom, and eat not a man nor slay him, but in great hunger. The Lion is in most gentleness and nobility, when his neck and shoulders be heled with hair and mane; and he that is gendered of the pard lacketh that nobility. The Lion knoweth by smell if the pard gendereth with the lioness, and reseth [rageth] against the lioness that breaketh spousehead [wedlock], and punisheth her full sore, but if she wash her in a river, and then it is not known to the Lion. And when the lioness whelpeth, her womb is rent with the claws of her whelps, and whelpeth therefore not oft. And the lioness whelpeth first five whelps, and afterward four, and so each year less by one, and waxeth barren when she whelpeth one at last. And she whelpeth whelps evil shapen and small, in quantity [*i.e.*, size] of a weasel in the beginning. And whelps of six months may unneath [hardly] be whelped, and whelps of two months may unneath move. And when the Lion eateth once enough, afterward he is meatless two days or three. And if him needeth to flee, he casteth up his meat into his mouth, and draweth it out with his claws, to be in that wise the more light to run and to flee. The Lion liveth most long, and that is known by working and wasting of his teeth; and then in age he reseth on a man; for his virtue and might faileth to pursue great beasts and wild. And then he besiegeth cities to ransom, and to take men; but when the Lions be taken, then they be hanged, for other Lions should dread such manner pain. The old Lion reseth woodly [madly, furiously] on men, and only grunteth on women, and reseth seld on children, but in great hunger. By the tail the boldness and heart of the Lion is known, as the horse is known by the ears. For when the Lion is wroth, first he beateth the earth with his tail, and afterward, as the wrath increaseth, he smiteth and beateth his own back. And out of each wound, that the Lion maketh with claw or with teeth, runneth sharp and sour blood. Also in peril the Lion is most gentle and noble, for when he is pursued with hounds and with hunters, the Lion lurketh not nor hideth himself, but sitteth in fields, where he may be seen, and arrayeth himself to defence. And he hideth himself not for dread that he hath, but he dreadeth himself sometime, only for he would not be dread. When he is wounded, he taketh wonderly heed, and knoweth them that him first smiteth,

and reseth on the smiter, though he be in never so great multitude; and if a man shoot at him [and do not hit him—*Bartholomew*] the Lion chaseth him and throweth him down, and woundeth him not, nor hurteth him. When the Lion dieth, he biteth the earth, and tears fall out of his eyes; and when he is sick, he is healed and holpen with the blood of an ape. And he dreadeth greatly the crowing and the comb of a cock. And the Lion hath a neck as it were unmovable, and is full grim; and moveth alway first with the right foot, and afterward with the left foot, as a camel doth; and [hath] little marrow in his bones; and his bones be so hard that by smiting of them together, fire springeth out thereof. The Lion dreadeth when he seeth or heareth a whelp beaten. He hideth himself in high mountains, and espieth from thence his prey. And he maketh a circle all about other beasts with his tail, and all the beasts dread to pass out over the line of the circle, and the beasts stand astonied and afeard, as it were abiding the hest and commandment of their King. And he is ashamed to eat alone the prey that he taketh; therefore of his grace, of free heart, he leaveth some of his prey to other beasts that follow him afar off. And is so hot of his complexion, that he hath alway the fever quartan; and hath kindly this evil to abate his fierceness. His grease is contrary to venom, so that whoso be anointed therewith shall not dread that time biting of serpents nor creeping worms. Also his grease meddled with oil of roses keepeth and saveth the skin of the face from wens and vices, and keepeth whiteness. His gall meddled with water sharpeth and cleareth the sight, and helpeth against infecting evils. His heart taken in meat destroyeth the fever quartan.

Bartholomew (Berthelet), bk. xviii. § 65.

THE Lion has a strong smell, and especially in the mouth. When he sleeps in a ship, the ship is in danger. The Lion flees before a mouse, and is afraid of the wood which is called sethin. Hellebore too and squill kill dogs and Lions and many wild beasts.

Hortus Sanitatis, bk. ii. § 80.

THE circles of cart-wheels, empty carts, and the comb on a cock's head do marvellously fear a Lion being a most hardy or fierce beast.

Lupton, "Notable Things," bk. iii. § 37.

IF you join a Lion's skin to the skin of a wolf or any other beast, it will make them without hair, or cause their hair to fall or consume away. *Ibid.*, bk. vi. § 54.

CLOTHES wrapped in a Lion's skin killeth moths. And so great is the fear of Lions to wolves, that if any part of a Lion's grease be cast into a fountain, the wolves never dare to drink thereof, or to come near unto it. The flesh of a Lion being eaten either by a man or woman, which is troubled with dreams and fantasies in the night-time, will very speedily and effectually work him ease and quietness. The grease of a Lion being dissolved and presently again conglutinated together, and so being anointed upon the body of those who are heavy and sad, it will speedily extirpate all sorrow and grief from their hearts. The gall of a Lion being taken in drink by anyone doth kill or poison him out of hand.

Topsell, "Four-footed Beasts," pp. 376-9.

If thongs be cut from a Lion's skin, a man girt with them will not fear his foes; and if his eyes be put under the arm-pits, or worn, all beasts will bow their heads, and flee behind his back.

Albertus Magnus, "Of the Virtues of Animals."

Lions we have had very many in the North parts of Scotland, and those with manes of no less force than those of Mauritania; but how and when they were destroyed as yet I do not read.

Holinshed, "Description of England," p. 225.

In Pietra Rossa [Barbary], the Lions are so tame that they will gather up bones in the streets, the people not fearing them. The like Lions are in Guraigura, where one may drive them away with a staff. At Agla the Lions are so fearful, that they will flee at the voice of a child, whence a cowardly braggart is proverbially called a Lion of Agla. *Purchas'* "Pilgrims," p. 621 (ed. 1616).

Lioness.

As You Like It, iv. 3, 115.

The Lioness is more cruel than the lion, and namely when she hath whelps, for she putteth her in peril of death for her whelps. There is a little beast that the lion and the Lioness dreadeth wonderfully, and that beast hight *Leontophonus*. For that beast beareth a certain venom, which slayeth the lion and the Lioness. Therefore this said beast is taken, and afterward burnt, and the flesh [which] is sprung [sprinkled] with the ashes, and laid and set in meeting of ways, shall slay and destroy the lions which eat thereof. The lion's breath stinketh, and is right infectious and contagious, and infecteth other things, and his biting is deadly and venomous, and namely when he is wood. For the lion waxeth wood, as the hound doth. *Bartholomew* (*Berthelet*), bk. xviii. § 66.

STRANGE it is that a Lioness, by showing her hinder parts to the male, should make him run away.

Purchas' "Pilgrims," p. 557 (ed. 1616).

Lizard.

Venom toads, or lizards' dreadful stings.
iii. KING HENRY VI., ii. 2, 138.

THE Lizard is a little beast painted on the back with shining specks as it were stars. The Lizard is so contrary to scorpions that the scorpions dread and lose comfort when they see the Lizard. The Lizard liveth most by dew; and though he be a fair beast and fair painted, yet he is right venomous; for the worst medicine is made of the Lizard, for when he is dead in wine, he covereth their faces that drink thereof with vile scabs; therefore they eschew [? ensue. *Bartholomew* has—"for this ointment they who envy the fairness of strumpets kill the Lizard"] to put him in medicine and ointment that have envy to fairness of strumpets. His remedy is the yolk of an egg, honey and glass. And the gall of a Lizard stamped in water assembleth together weasels. And the Lizard lurketh in winter in dens and chines, and his sight dimmeth; and in springing time he cometh out of his den, and feeleth that his sight faileth, and changeth his place, and seeketh him a place toward the East, and openeth continually his eyes toward the rising of the sun, until the humour in the eye be full dried, and the mist wasted that is cause of dimness in the eye.

Bartholomew (*Berthelet*), bk. xviii. § 94.

A GREEN Lizard hath a great delight to behold a man in the face, for he will lovingly fawn upon him as a dog with the moving of his tail. And as much as in him lies, will defend him from a serpent that lies lurking in the neaths to hurt him.

Lupton, "Notable Things," bk. vi. § 73.

THE venom of the Lizard is deadly, and the remedy for it is made from the pounded flesh of scorpions. There is no animal more deceitful than the Lizard, and he envies man. In the flesh of the Lizard is virtue for extracting splinters and thorns. Its fat fattens much.

Hortus Sanitatis, bk. ii. § 130.

THE Lizard when he lies
Too open to the hot sun faints and dies.

Heywood's "Anna and Phillis," emb. 16.

WHEN a certain man had taken a great fat Lizard, he did put out her eyes with an instrument of brass, and so put her into a new earthen pot, which had in it two small holes or passages, big enough to take breath at, but too little to creep out at, and, with her, moist earth and a certain herb; and furthermore he took an iron ring, wherein was set an engagataes [? agate] stone with the picture of a Lizard engraven upon it; and besides upon the ring he made nine several marks, whereof he put out every day one, until at the last he came at the ninth, and then he opened the pot again, and the Lizard did see as perfectly as ever he did before the eyes were put out. The old one devoureth the young ones as soon as they be hatched, except one which she suffereth to live, and this one is the basest and most dullard; yet notwithstanding, afterwards it devoureth both his parents. Twice a year they change their skin. They live by couples together, and when one of them is taken, the other waxeth mad, and rageth upon him that took it. They are enemies to bees. They fight with all kind of serpents. The eggs of Lizards do kill speedily, except there come a remedy from falcon's dung and pure wine. Mingled with oil it causeth hair to grow again upon the head of a man. If green Lizards see a man, they instantly gather about him, and laying their heads at the one side with great admiration behold his face. The use of these green Lizards is by their skin and gall to keep apples from rotting, and also to drive away caterpillars, by hanging up the skin on the tops of trees, and by touching the apples with the said gall. The ashes of a green Lizard do reduce scars in the body to their own colour.

Topsell, "History of Serpents," pp. 739-42.

TAKE a Lizard and cut off its tail, and take what comes out, because it is like quicksilver. Then take a taper, and moisten it with oil, and put it in a new lamp, and light it,—that man's house will appear splendid and white or silvered.

Albertus Magnus, "Of the Wonders of the World."

[IN the deserts of Lybia is] a kind of great Lizard which never drinketh, and, if water be put in his mouth, he presently dieth. *Purchas'* "Pilgrims," p. 559 (ed. 1616).

Loach.

i. KING HENRY IV., ii. 1, 23.

THE Loach is a little river-fish, white with black spots. Some say that it feeds on dead bodies, but this is held by fishermen to be fabulous. They are considered poor and contemptible eating. *Hortus Sanitatis*, bk. iv. § 41.

[The commentators are puzzled by this passage in "i. King Henry IV." It is quite probable that "like a Loach" has no more accurate meaning than "like a house afire," or "like the deuce." But "Some fishes there be, which of themselves are given to breed fleas and lice, among which the *chalcis*, a kind of turbot, is one" (*Philemon Holland's Pliny*, bk. ix. ch. xlvii.)—and perhaps the Loach is another.]

Locust.

OTHELLO, i. 3, 354.

[May perhaps be the fruit of the Locust-tree.]

LOCUST hath that name for it hath long legs as the shaft of a spear. And these worms that hight Locust have no king, and yet they pass forth ordinately in companies. And hath a square mouth, and a sting instead of a tail, and crooked and folding legs. And are gendered of the southern wind, and excited to flight. And they die in the Northern wind. Also this worm Locust for the most part is all womb, and therefore it hath never meat enough. And of their dirt worms be gendered.

Bartholomew (*Berthelet*), bk. xii. ("Of Birds") § 24.

HE burneth corn with touching, and devoureth the residue. In India be of them three foot in length, which the people of the country do eat.

Batman's addition, *ut supra.*

THE Locust [is] none other creature than the grass-hopper. In Barbary [etc.] they are eaten; nevertheless they shorten the life of the eaters by the production at the last of an irksome and filthy disease. In India they are three foot long, in Ethiopia much shorter.

Holinshed, "Description of England," p. 229.

Louse.

TROILUS AND CRESSIDA, v. 1, 72.

A LOUSE is a worm of the skin, and grieveth more in the skin with the feet and with creeping, than he doth with biting, and is gendered of right corrupt air and vapours, that sweat out between the skin and the flesh by pores. And some lice gender of sanguine humour, and be red and great; and some of phlegmatic humours, and they be nesh and white; and some of choleric humours, and be citrine [*i.e.*, yellow], long, swift and sharp; some of melancholic humour, and they be coloured as ashes, and be lean and slow in moving. And the leaner that a Louse is, the sharper she biteth and grieveth.

Bartholomew (*Berthelet*), bk. xviii. § 88.

LICE cometh also of that cloth that is trained in the wool with the fat or grease of an horse or of a swine, and therefore the Northern cloths worn of a sweating body do breed lice in 12 hours.

Batman's addition to *Bartholomew*, bk. xviii. § 116.

THIS disease is undoubtedly created from the very flesh of man, and yet invisibly. *Hortus Sanitatis*, bk. ii. § 119.

THE old skin or slough that snakes do cast off in the spring, whosoever drinketh in his ordinary drink, it will kill all the vermin or Lice of the body within three days.

Holland's Pliny, bk. xxx. ch. xv.

[So many remedies are given for this complaint, that it must have been very common.]

Our doublets were lined with taffeta, wherein Lice cannot breed or harbour; so as howsoever I wore one and the same doublet till my return into England, yet I found not the least uncleanliness therein.

Fynes Moryson, "Itinerary," part i. p. 209.

V. Flea.

Love-in-Idleness.

Midsummer Night's Dream, ii. 1, 168.

V. Heart's-ease, Pansy.

Luce.

Merry Wives of Windsor, i. 1, 16.

The Luce feeds on poisons, toads and such like; yet it is said to be good food for the sick. If the net in which it has been caught be lifted from the water so that it sees the light of day, it rarely or never happens that it remains any longer, but seeks itself some way out. The Luce has in its brain a stone like crystal, but only when it has lived long. *Hortus Sanitatis*, bk. iv. § 53.

[The Sea-Luce is the codfish.]

Mace.

Winter's Tale, iv. 3, 49.

V. Nutmeg.

Mackerel.

i. King Henry IV., ii. 4, 395.

When Mackerel ceaseth from the seas
John Baptist brings grass-beef and pease.

Tusser, "Good Husbandry:" "The Farmer's Daily Diet."

This law-French is worse than butter'd Mackerel,—
Full o' bones, full o' bones.

"Life and Death of Captain Thos. Stucley," line 291.

THE old tunnies and the young enter into the sea Pontus; and every company of them hath their several leaders and captains; and before them all the Mackerels lead the way, which, while they be in the water, have a colour of brimstone; but without, like they be to the rest.

Holland's Pliny, bk. ix. ch. xv.

Maggot-pie.

MACBETH, iii. 4, 125.

MAGOTAPIE.
Minsheu's Dictionary.

THE Magpie makes up for the shortness of its wings by the length of its tail. It builds its nest with two holes,—by the one it goes in, and at the other it puts out its tail.

Hortus Sanitatis, bk. iii. § 99.

PIES take a love to the words that they speak; for they not only learn them as a lesson, but they learn them with a delight and pleasure; and by their careful thinking upon that which they learn, they shew plainly how mindful and intentive they be thereto. It is for certain known that they have died for very anger and grief that they could not learn to pronounce some hard words; as also, unless they hear the same words repeated often unto them, their memory is so shittle, they will soon forget the same again.

Holland's Pliny, bk. x. ch. xlii.

IF you wish to loose chains, go into a wood, and look where a Pie has her nest with young ones, and when you are there, climb up the tree, and bind the opening [of the nest] round with anything you please, because, when the Pie sees you, it goes for a certain herb, which it puts to the bond, and forthwith it is broken, and then that herb falls to the ground onto a cloth which you should have put under the tree, and do you be handy and take it.

Albertus Magnus, "Of the Wonders of the World."

Mallard.

ANTONY AND CLEOPATRA, iii. 10, 20.

THE blood of ducks and Mallards bred in the realm of Pontus is passing good for any such indirect means wrought by poison or witchcraft; and therefore their blood is

ordinarily kept dry in a thick mass, and as need requireth is dissolved and given in wine; but some think that the blood of the female duck is better than that of the Mallard or drake.
Holland's Pliny, bk. xxix. ch. v.

YOUR citizens' wives love green geese in spring, Mallard and teal in the fall, and woodcock in winter.
Webster, "Westward Ho!" i. 1.

[Mallards were boiled with cabbage or onions ("Good Huswife's Handmaid," p. 5), or stewed ("The Good Huswife's Jewel").]

Mallow.

TEMPEST, ii. 1, 144.

HE that is balmed with the juice of the hock [*i.e.*, Mallow—hollyhock], meddled with oil may not be grieved with stinging of bees. Also members balmed with juice thereof be not bitten of attercops [*i.e.*, spiders], nor stung of scorpions. The broth thereof maketh sleep, if the face be washed therewith, and the outer parts of the body.
Bartholomew (*Berthelet*), bk. xvii. § 107.

IF a man or woman sup off a small draught (though it were no more but half a spoonful) every day of the juice of any Mallow, it skills not which, he shall be free from all diseases and live in perfect health.
Holland's Pliny, bk. xx. ch. xxi.

[Mallows were eaten as a vegetable (*cf.* "The Good Huswife's Handmaid," p. 1, and *Evelyn's* "Acetaria," § 40).]

Malmsey.

KING RICHARD III., i. 4, 161.
ii. KING HENRY IV., ii. 1, 42, etc.

MALMSEY and muscadine were wines of Candia.
Fynes Moryson, "Itinerary," part iii.

THE Vintners of the Low Countries (I will not say of London) do make of Cute and wine mixed in a certain proportion, a compound and counterfeit wine, which they sell for Candy wine, commonly called Malmsey.

Gerard's "Herbal," bk. ii. ch. cccxxiii.

I LOVE thee next to Malmsey in a morning.

Beaumont and Fletcher, "The Captain," iv. 2.

Malt.

KING LEAR, iii. 2, 82.

OUR Malt is made all the year long in some great towns, but in gentlemen's and yeomen's houses, who commonly make sufficient for their own expenses only, the winter half is thought most meet for that commodity; the Malt that is made when the willow doth bud, is commonly worst of all. The best Malt is tried by the hardness and colour, for if it look fresh with a yellow hue, and thereto will write like a piece of chalk, after you have bitten a kernel in sunder in the midst, then you may assure yourself that it is dried down—of all, the straw-dried is the most excellent. For the wood-dried Malt, when it is brewed, doth hurt and annoy the head of him that is not used thereto. *Holinshed*, "Description of England," p. 169.

Mandragora, Mandrake.

Not poppy nor mandragora
Nor all the drowsy syrups of the world
Shall ever medicine thee to that sweet sleep
Which thou owedst yesterday.

OTHELLO, iii. 3, 330.

Kill, as doth the mandrake's groan.

ii. KING HENRY VI., iii. 2, 310.

MANDRAGORA beareth apples with great savour. The rind thereof meddled with wine is given to them to drink that shall be cut in the body, for they should sleep and not feel the sore cutting. And apples grow on the leaves, and be yellow and sweet of smell, but with a manner heaviness, and be fresh in savour. But yet Mandragora must be warily used, for it slayeth if men take much thereof. The juice thereof with woman's milk laid to the temples maketh to sleep, yea though it were in the most hot ague. Mandragora hath many other virtues, and smiteth off and destroyeth swelling of the body, and withstandeth venomous biting. They that dig Mandragora be busy to beware of contrary winds, while they dig, and make three circles about with a sword, and abide with the digging unto the sun going down, and trow so to have the herb with the chief virtues. The juice thereof is gathered and dried in the sun, the apples thereof be dried in the shadow.

Bartholomew (*Berthelet*), bk. xvii. § 104.

THERE hath been many ridiculous tales brought up of this plant, whether of old wives, or of some runagate surgeons or physic-mongers I know not. That it is never or very seldom to be found growing naturally but under a gallows, where the matter that hath fallen from a dead body hath given it the shape of a man; and the matter of a woman the substance of a female plant. That he who would take up a plant thereof must tie a dog thereunto to pull it up, which will give a great shriek at the digging up; otherwise if a man should do it, he should surely die in short space after. *Gerard's* "Herbal," *s.v.*

[Full directions are given by Lupton ("Notable Things," bk. iii. § 43) "to make the counterfeit Mandrake, which hath been sold by deceivers for much money," from the "great double root of Briony"—hair to be imitated by millet-seeds—and the whole to be buried "until it have gotten upon it a certain little skin."]

A MANDRAKE'S voice, whose tunes are cries
So piercing that the hearer dies.

Dekker, "Double P. P."

WITHOUT the death of some living thing it cannot be drawn out of the earth to man's use. Therefore they did tie some dog or some other living beast unto the root thereof with a cord, and digged the earth in compass round about, and in the mean time stopped their own ears for fear of the terrible shriek and cry of this Mandrake. In which cry it doth not only die itself, but the fear thereof killeth the dog or beast which pulleth it out of the earth.

Bullein, "Bulwark of Defence against Sickness," p. 41 quoted in *Reed's* Shakespeare.

If the root be seethed for six hours with ivory, it softens it and makes it easy to work into any shape desired. *Hortus Sanitatis*, bk. i. § 276.

I FRAMED a mole under my child's ear by art; you shall see it taken away with the juice of Mandrake.

Lilly, "Mother Bombie," v. 3.

Marble.

MACBETH, iii. 4, 22.

MARBLE STONES be noble stones, and be praised for speckles and diverse colours. Over all things we may wonder that Marble stones be not hewed neither cloven with iron, neither with steel, with hammer nor with saw, as they be with a plate of lead set between nesh [soft] shingles or spoons. For with lead and not with iron Marble stones be hewen and cloven and planed, as shingles or small stones. *Bartholomew* (*Berthelet*), bk. xvi. § 69.

MANY mines of coarse and fine Marble are there in England; but chiefly one in Staffordshire, another near to the Peak, the third at Vauldry (?), the fourth at Snothill (?) (belonging to the Lord Chandos), the fifth at Eaglestone, which is of black Marble spotted with grey or white spots, the sixth not far from Durham. Of white Marble also we have store. The black Marble spotted with green is none of the vilest sort.

Holinshed, "Description of England," p. 235.

Mare.

KING HENRY V., ii. 1, 25.

THE name of an horse's wife shall be called a Mare. And if a Mare, being with foal, smelleth the snuff of a candle, she casteth her foal. Also in the forehead of the colt breedeth a black skin of the quantity of a sedge, and the mother licketh it with her tongue, and taketh it away, and receiveth never the colt to suck her teats, but it be first taken away. Also the Mare is proud, and hath joy of her mane, and is sorry when it is shorn, as though the virtue of love were in the mane. Also a bird that hight Ibis [the stork] fighteth with the horse, because the horse driveth her out of her pasture and leys; for the stork is feeble of sight, and hath a voice as an horse; and when he flieth above an horse, he stonieth [astonisheth] him, and maketh him flee, and slayeth him sometime.

Bartholomew (*Berthelet*), bk. xviii. § 40.

A MARE will bring forth a foal of divers colours if she be covered with a cloth of divers colours, whiles she is taking the horse. The same may be proved with dogs and other beasts. *Lupton*, "Notable Things," bk. v. § 1.

Marigold.

The marigold that goes to bed wi' the sun
And with him rises weeping.

WINTER'S TALE, iv. 4, 105.

[Marigold is one of the herbs to make broth and farcing (enumerated in the "History of Jacob and Esau," iv. 5).]

I THINK of kings' favours as of a Marigold flower
That, as long as the sun shineth, openeth her leaves,
And with the least cloud closeth again.

"A Knack to Know a Knave."

IF the mouth be washed with the juice, it helpeth the tooth-ache. The yellow leaves of the flowers are dried and kept throughout Dutchland against winter to put into broths, in physical potions, and for divers others purposes, in such quantity, that in some grocers' or spice-sellers' houses are to be found barrels filled with them, and retailed by the penny more or less, insomuch that no broths are well made without dry Marigolds.

Gerard's "Herbal," *s.v.*

Lupton ("A Thousand Notable Things," bk. vi. § 85, and bk. iv. § 79) confuses Marigolds with sunflowers.]

THUS the Marigold opens at the splendour of a hot constant friendship. 'twixt you both.

Middleton, "No Wit, No Help Like a Woman's," i. 1.

Marjoram.

WINTER'S TALE, iv. 4, 104.

SWEET Marjoram is a remedy against cold diseases of the brain and head, being taken any way to your best liking; put up into the nostrils it provoketh sneezing, and draweth forth much baggage phlegm; it easeth the toothache being chewed in the mouth; being drunk, it is used in medicines against poison. The leaves, dried and mingled with honey, and given, dissolveth congealed or clotted blood, and putteth away black and blue marks after stripes and bruises being applied thereto. *Gerard's* "Herbal," *s.v.*

As a plaster or drink, it cures those grieved by a scorpion, if mixed with vinegar and salt.

Hortus Sanitatis, bk. i. § 409.

WILD Marjoram or Organy is profitably used in a looch, or medicine to be licked, against an old cough and the stuffing of the lungs. The herb strewed upon the ground driveth away serpents. *Gerard's* "Herbal," *s.v.*

Marl.

MUCH ADO ABOUT NOTHING, ii. 1, 66.

WE have pits of fat and white and other-coloured Marl, wherewith in many places the inhabiters do compost their soil, and which doth benefit their land in ample manner for many years to come.

Holinshed, "Description of England," p. 236.

WE have a kind of white Marl, which is of so great force, that if it be cast over a piece of land but once in three score years, it shall not need of any further composting. [It] lieth sometime a hundred foot deep.

Ibid., p. 109.

Marmoset.

TEMPEST, ii. 2, 174.

HUSBAND is like your clog to your Marmoset.

Ben Jonson's "Poetaster," iv. 2.

V. Monkey.

Martlet.

The martlet
Builds in the weather on the outward wall.

MERCHANT OF VENICE, ii. 9, 28.

v. MACBETH, i. 6, 3-10.

MARTINETS, Martins, Martlets. — These birds are so called because they come to us about the end of the month of March from warm regions, and depart before the feast of St. Martin.

Minsheu's Dictionary, *s.v.*

MARTINS are good to eat.

Batman's addition to *Bartholomew*, bk. xii. § 21.

V. Swallow.

Mary-bud.

CYMBELINE, ii. 3, 25.

V. Marigold.

Mast.

The oaks bear mast.

TIMON OF ATHENS, iv. 3, 422.

THE oak bringeth forth a profitable kind of Mast, whereby such as dwell near unto the aforesaid places do cherish and bring up innumerable herds of swine. [Red and fallow deer eat Mast also]; yea, our common poultry if they may come unto them; but those eggs which these latter do bring forth (beside blackness in colour and bitterness of taste) have not seldom been found to breed divers diseases unto such persons as have eaten of the same. (The like have I seen where hens do feed upon the tender blades of garlick [marginal note].)

Holinshed, "Description of England," p. 214.

Mastiff.

KING HENRY V., iii. 7, 151.

OR Bandog—keeping the house—or *molossus*.

Minsheu's Dictionary, *s.v.*

Medlar.

AS YOU LIKE IT, iii. 2, 125.
MEASURE FOR MEASURE, iv. 3, 184.
ROMEO AND JULIET, ii. 1, 36.
So also *Middleton*, "Women Beware Women," iv. 2, 100.

[The prejudices of the nineteenth century will not allow the popular name of the Medlar in the sixteenth and seventeenth centuries to be recalled.]

IF you graft the slips of a wilding or sour apple upon the stock of a hawthorn, you shall have Medlars grow thereof. This I have seen proved: therefore I affirm it for a very truth. *Lupton*, "Notable Things," bk. x. § 95.

THE graft should be taken from the middle of the tree; for one from the top is faulty.

Hortus Sanitatis, bk. i. § 294.

THE fruit of the three-grain [Neapolitan] Medlar is eaten both raw and boiled. These Medlars be oftentimes preserved with sugar or honey. Moreover, they are singular good for women with child; for they strengthen the stomach, and stay the loathsomeness thereof.

Gerard's "Herbal," *s.v.*

MEDLARS, the best:—The Great Dutch, Neapolitan, and one without stones.

Evelyn, "Kalendarium Hortense," p. 274.

Mermaid.

I'll drown more sailors than the mermaid shall.
iii. KING HENRY VI., iii. 2, 186.

V. **Siren.**

Milk.

WHEN cow's Milk is first congealed, it is as it were a stone, and that happeth when it is meddled with water. Also when a child is nourished with hot Milk, his teeth springeth the sooner. And if any hair cometh therein, there falleth a great sickness; and the ache ceaseth not ere the hair cometh out with the Milk, or rotteth. And a black woman hath much better Milk and more nourishing than a white woman. A drop of good Milk put on the nail abideth continually, and droppeth not away.

Bartholomew (*Berthelet*), bk. xix. § 63.

Cow's Milk is the better and more wholesome, if the most deal of wateriness be consumed and wasted by stones of the rivers that be heated fiery hot and then quenched therein.

Ibid., § 65.

[Stow, in his "Survey of London," gives the price of milk in his youth (*circa* 1535) as three ale-pints for a halfpenny in the summer, nor less than one ale-quart for a halfpenny in the winter, always hot from the kine, and he fetched it from the Minories farm just outside Aldgate.]

Mint.

WINTER'S TALE, iv. 4, 104.

MINT is an herb with good smell, and thereof is double kind, wild and tame. It taketh away abomination of wambling, and abateth the yexing [*i.e.*, hiccough].

Bartholomew (*Berthelet*), bk. xvii. § 106.

IT is taken inwardly against scolopenders, bear-worms, sea-scorpions and serpents. It is applied with salt to the bitings of mad dogs. It will not suffer milk to curdle in the stomach, therefore it is put in milk that is drunk for fear that those who have drunk thereof should be strangled.

Gerard's "Herbal," *s.v.*

Misletoe.

TITUS ANDRONICUS, ii. 3, 95.

MISTLETOE with red lily opens all locks. If the aforesaid be hung on a tree with the wing of a swallow, thither will congregate all the birds within quite five miles, and this last has been tried in my time.

Albertus Magnus, "Of Virtues of Herbs," § 10.

Mite.

ALL'S WELL THAT ENDS WELL, i. 1, 154.

V. Worm.

Moldwarp.

i. KING HENRY IV., iii. 1, 149.

Mole.

TEMPEST, iv. 1, 194.

A MOLE is a little beast somewhat like unto a mouse. And he is damned in everlasting blindness and darkness, and is without eyes, and hath a snout as a swine, and diggeth therewith the earth, and casteth up that he diggeth, and gnaweth and eateth roots under earth, and hateth the sun, and may not live above the earth. And the Mole hath none eyes seen without, and who that slitteth the skin subtly and warily shall find within the fores [*i.e.*, traces] of eyes hidden. And some men trow that that skin breaketh for anguish and for sorrow when he beginneth to die, and beginneth then to open the eyes in dying that were closed living. *Bartholomew* (*Berthelet*), bk. xviii. § 102.

A MOLE or want enclosed in an earthen pot, if you set then the powder of brimstone on fire, she will call other Moles or wants to help her with a very mourning voice.

Lupton, "Notable Things," bk. iii. § 6.

If you will catch Moles or wants, put garlic, leeks or an onion in the mouths of their holes, and you shall see them come or leap out quickly, as though they were amazed or astonied. *Bartholomew* (*Berthelet*), bk. ix. § 14.

Mole's blood sprinkled on a bald head makes the hairs come back. *Hortus Sanitatis*, bk. ii. § 139.

Moles have no ears, and yet they understand all speeches spoken of themselves. If a man eat the heart of a Mole newly taken out of her belly and panting, he shall be able to divine and foretell infallible events. There is nothing which is more profitable or medicinable for the curing of the bites of a shrew, than a Mole being flayed and clapped thereunto. For the changing of the hairs of horses from black to white, take a Mole and boil her in salt water, or lye made of ashes three days together, and when the water or lye shall be quite consumed, put new water or lye thereunto; this being done, wash or bathe the place with the water or lye somewhat hot; presently the black hairs will fall and slide away, and in some short time there will come white. *Topsell*, "Four-footed Beasts," pp. 389-91.

If the foot of a Mole be wrapped in a laurel-leaf and put into a horse's ear, he will run away for fear; and if it be put in the nest of any bird, no young ones will be hatched out of those eggs.

Albertus Magnus, "Of the Virtues of Animals."

Monkey.

Merchant of Venice, iii. 1, 124.

V. Ape.

[Monkeys were common pets for ladies, and shared their favour with dogs and paroquets (so *Massinger*, "New Way to Pay Old Debts"; *Ben Jonson*, "Cynthia's Revels"; *Middleton*, "Michaelmas Term," etc.).]

Moon-calf.

Tempest, ii. 2, 111.

A false conception called *mola* or Moon-calf, that is to say, a lump of flesh without shape, without life, and so hard withal, that unneath a knife will enter and pierce it either with edge or point. Howbeit a kind of moving it hath.

Holland's Pliny, bk. vii. ch. xv.

Moss.

TITUS ANDRONICUS, ii. 3, 95.

[Gerard (" Herbal," *s.v.*) states that he has found "goldilocks, or golden maiden-hair, the bigger and less in great abundance," between Hampstead Heath and Highgate, and "club-moss or wolf-claw Moss" only upon Hampstead Heath, and he testifies that the latter Moss, if it be hanged in a vessel of "floating wine, which is now become slimy," the wine is restored to his former goodness. Moreover, the kind of Moss, which "is found upon the skulls or bare scalps of men and women lying long in charnel-houses or other places, where the bones of men and women are kept together, is thought to be a singular remedy against the falling evil, and the chin-cough in children, if it be produced and then given in sweet wine for certain days together."]

Moth.

CORIOLANUS, i. 3, 94.

A MOTH is a worm of clothes, and is gendered of corruption of cloth, when the cloth is too long in press and thick air, and is not blown with wind, neither unfolded in pure air. And though he be a sensible beast, yet he hideth himself within the cloth, that unneath he is seen. Leaves of the laurel-tree, of cedars, and of cypress, and other such, put among clothes in hutches save the clothes and also books from corruption and eating of Moths.

Bartholomew (*Berthelet*), bk. xviii. § 105.

IF you seethe the dregs, or mother, or foam of oil to the half, and therewith anoint the bottom, corners, and feet of any chest or press,—the clothes that you lay therein shall never be hurt with Moths (so that it be dry before you put therein your clothes).

Lupton, "Notable Things," bk. ii. § 94.

MOTHS breed in garments so much the sooner if a spider be shut in. They that sell woollen clothes use to wrap up the skin of a bird called the king-fisher amongst them, or else hang one in the shop, as a thing by a secret antipathy that Moths cannot endure. Garments wrapped up in a lion's skin will never have any moths.

Mouffet, "Theatre of Insects," p. 1100.

Mouse.

MIDSUMMER NIGHT'S DREAM, V. I, 394.

THE Mouse is a little beast, and he breedeth and is gendered of humours of the earth. Also the liver of this beast waxeth in the full of the moon, like as a certain fish of the sea increaseth then, and waneth again in the waning of the moon. And the Mouse drinketh not, and if he drinketh, he dieth; and is a gluttonous beast, and is therefore beguiled with a little meat when he smelleth it, and will taste thereof. His urine stinketh and is contagious; and his biting is venomous, and also his tail is venomous accounted. In harvest the male and female gather corn, and charge either other upon the womb, and the male draweth the female so charged by the tail to her den, and dischargeth her, and layeth up that stuff in a place in the den, and thus they go again to travail, and gather ears of corn, and the male layeth himself on his own back, and his female chargeth him, and taketh his tail in her mouth, and draweth him so home to the den. And though mice be full grievous and noyful beasts, yet they be in many things good and profitable in medicine. Mice dirt bruised with vinegar keepeth and saveth the head from falling of hair. His new skin laid all about the heel, healeth and saveth kibes and wounds therefro.

Bartholomew (*Berthelet*), bk. xviii. § 73.

PUT one or more quick [*i.e.*, live] Mice in a long or deep earthen pot, and set the same unto a fire made of ash-wood; and when the pot begins to wax hot, the Mice therein will chirp or make a noise; whereat all the mice that are nigh them will run towards them, and so will leap into the fire, as though they should come to help their poor imprisoned friends and neighbours. The cause whereof Mizaldus ascribes to the smoke of the ash-wood.

Lupton, "Notable Things," bk. x. § 93.

Writing-ink tempered with water, wine, or vinegar, wherein wormwood hath been steeped, Mice will not eat the papers or letters written with that ink.

Lupton, "Notable Things," bk. vi. § 51.

Quicksilver killed, burnt lead, the scales of iron, or black hellebore, mixed with some pleasant meat that Mice love—if any Mice eat thereof, it will kill them.

Ibid., bk. x. § 68.

Of the vulgar little Mouse :—Concerning their manners they are evil, apt to steal, insidious and deceitful. If the brains of a weasel, the hair or rennet be sprinkled upon cheese, or any other meat whereunto Mice resort, they not only forbear to eat thereof, but also to come in that place. A Mouse watcheth an oyster when he gapeth, and seeing it open, thrusts in his head to eat the fish ; as soon as ever the oyster felt his teeth, presently he closeth his shell again, and so crusheth the Mouse's head in pieces. A man took a Mouse, which Mouse he fed only with the flesh of Mice, and after he had fed it so a long time, he let it go, who killed all the Mice he did meet, and was not satisfied with them, but went into every hole that he could find, and ate them up also. A Mouse being flayed, and afterwards cut through the middle, and put unto a wound or sore wherein there is the head of a dart or arrow, or any other thing whatsoever, will presently and very easily exhale and draw them out of the same. A young Mouse mingled with salt is an excellent remedy against the biting of the Mouse called the shrew, which biting horses and labouring cattle, it doth venom until it come unto the heart, and then they die, except the aforesaid remedy be used. Of the heads of Mice being burned is made that excellent powder for the scouring and cleansing of the teeth called tooth-soap. For the rottenness and diminishing of the teeth, the best remedy is to take a living Mouse, and to take out one of her teeth, whether the greatest or the least is no great matter, and hang it by the teeth of the party grieved ; but first kill the Mouse from whom you had the tooth, and he shall presently have ease and help of his pain.

Topsell, "Four-footed Beasts," pp. 392-402.

MICE are multiplied in dry seasons (which the store of them this dry winter—1613—confirmeth) of which there are great ones in Egypt with two feet which they use as hands, not going but leaping. *Purchas'* "Pilgrims," p. 560 (ed. 1616).

Mulberry.

MIDSUMMER NIGHT'S DREAM, iii. 1, 170.
CORIOLANUS, iii. 2, 79.

LEAVES thereof slayeth serpents, if they be thrown or laid upon them. The leaves sod in rain-water maketh black hair, and healeth the biting of attercops [spiders], and easeth the tooth-ache. Of Mulberries is noble drink made; elephants drink thereof, and be the more bold and hardy. *Bartholomew* (*Berthelet*), bk. xvii. § 100.

THE Mulberries [in Hegesander's time] did not bring forth fruit in twenty years together, and so great a plague of the gout then reigned and raged so generally, as not only men, but boys, wenches, eunuchs and women were troubled with that disease. *Gerard's* "Herbal," *s.v.*

Mule.

WINE-drinking is forbidden the Mule. The more water that the Mule drinketh, the more good his meat doth him. Also the Mule hath no gall openly seen upon his liver. *Bartholomew* (*Berthelet*), bk. xviii. § 72.

IF you fumigate a house with the left hoof of a Mule, no rat will remain in that house. The ashes of a Mule's hoofs cure baldness. *Hortus Sanitatis*, bk. ii. § 98.

MULES are broken of their flinging and wincing, if they use often to drink wine. *Holland's Pliny*, bk. viii. ch. xliv.

THE epithets of a Mule are these:—pack-bearer, dirty, Spanish, rough and bi-formed.

Topsell, "History of Four-footed Beasts," *s.v.*

Muscadel, or Muscadine.

TAMING OF THE SHREW, iii. 2, 172.

[Muscadel and brawn were usual refreshments at Christmas (so *Beaumont and Fletcher's* "Loyal Subject," iii. 4, also "Tamer Tamed," iv. 1, and "The Pilgrim," ii. 1.

Eggs and Muscadine were supposed to be restorative of the vital powers ("Tamer Tamed," i. 1; "Cupid's Revenge," i. 1; and many other plays of Massinger, Middleton, Brome, etc.).

Muscadines were also compounds to sweeten the breath (*Wardes'* "Treatise of Alexis of Piedmont's Secrets," 1562).]

MUSCADINES of Candia whereof and especially of red Muscadine there is great plenty in this island, wherewith England for the most part is served.

Fynes Moryson, "Itinerary," part i. p. 256.

Mushroom.

TEMPEST, v. 1, 39.

[Gerard describes not very clearly various kinds of edible and poisonous fungi, but thinks Mushrooms poor food.]

V. Toad-stool.

ITALIAN delicate oiled Mushrooms.

Massinger, "Guardian," ii. 2; so *Ben Jonson*, "Alchemist," ii. 2.

Two small casks—one of blue figs, the other of pickled Mushrooms. *Jasper Mayne*, "The City Match," v. 4.

Musk, Musk-cat.

MERRY WIVES OF WINDSOR, ii. 2, 68.
ALL'S WELL THAT ENDS WELL, v. 2, 21.

IN the mountains of Ind be some *Caprioli* [deer] that eateth herbs with good smell and savour, and in their feet be certain hollowness, in the which certain humours be gathered, and breedeth posthumes [*i.e.*, imposthumes, abscesses], the which posthumes first be ripened, and then broken with moving and with froting [*i.e.*, rubbing], and thrown out of the body with small hairy leaves. And the substance, that is contained within the skin, is best of smelling, and most precious among spicery, and most profitable and virtuous in medicine, and that we call commonly Musk. *Bartholomew* (*Berthelet*), bk. xviii. § 23.

IN the flank of the Musk-cat grows an imposthume from collected humours, and when this is ripe, the beast bruises and rubs it against a tree, and so it is broken, and the matter runs out, and thickens and hardens there, and the substance of the humour is called Musk. The whole of its flesh and its dung is named Musk, but that is far better which runs from the imposthume. Musk which has lost its smell recovers its virtue in stinks and latrines; it strives against stench, and thus revives as it were by striving.

Hortus Sanitatis, bk. ii. § 100.

THE Musk-cat is neither like a cat nor a mouse. They make perfume of it; with this the luxurious women perfume themselves, to entrap the love of their wooers. The true Musk is sold for 40s. an ounce at the least. A Musk-cat is an excellent remedy for those which are troubled with fear in their heart.

Topsell, "History of Four-footed Beasts," pp. 427-31 (1658).

[IN the Kingdom of Erginul] is found a beast, as big as a goat, of exquisite shape, which every full moon hath an aposthumation or swelling under the belly, which proveth the best Musk in the world.

Purchas' "Pilgrims," p. 428 (ed. 1616).

MUSK is made of the stomach of a beast somewhat greater than a cat. Our greatest sweet we see is but rottenness and putrefaction.

Ibid., p. 502.

V. Civet.

Musk-rose.

MIDSUMMER NIGHT'S DREAM, ii. 1, 252.

[The Musk-rose was a moss-rose. Gerard describes and engraves several species.]

Mussel.

TEMPEST, i. 2, 462.

THE Mussel is the male of the whale (*q.v.*). There are Mussels which are shell-fish, and from their milk oysters breed. The Mussel and the whale are examples of friendship, for as the whale's eyes through the great weight of its brows are closed, the Mussel swims before it and points out those things which might be harmful to its bulk, and the Mussel takes the place of eyes for the whale. This sea-mussel which precedes the whale has no teeth, but bristles instead. *Hortus Sanitatis*, bk. iv. § 57.

[Two distinct fishes are evidently here described, but the habits of both are interesting. The engraving is of a shell-fish like a whelk.

Mussels were seethed or boiled in their shells (second part of "The Good Huswife's Jewel," p. 53).]

Mustard.

TAMING OF THE SHREW, iv. 3, 22.

[Eaten with pancakes (AS YOU LIKE IT, i. 2, 66).]

SENVEY hight *Sinapis* [*i.e.*, Mustard], and healeth smiting of serpents and of scorpions, and overcometh venom of the scorpions, and abateth tooth-ache, and cleanseth the hair, and letteth the falling thereof. Bees love best the flowers and haunt them, and nevertheless bees touch never flowers of olive. *Bartholomew* (*Berthelet*), bk. xvii. § 155.

IF it be drunk fasting, it makes the intellect good.

Hortus Sanitatis, bk. i. § 436.

IT helpeth those that have their hair pulled off; it taketh away the blue and black marks that come of bruisings. *Gerard's* "Herbal," *s.v.*

[Tewkesbury was famous for Mustard ("ii. King Henry IV.," ii. 4, 262, and *Fynes Moryson*, "Itinerary," part iii. p. 139).]

SENVEY bruised and ground with vinegar is a wholesome sauce, meet to be eaten with hard and gross meats, either flesh or fish.

Batman's addition to *Bartholomew*, *l.c.*

Myrtle.

MEASURE FOR MEASURE, ii. 2, 117.

MYRTLE helpeth against venom, and against stinging of scorpions, if it be drunk; broth thereof helpeth against the falling of hair. Myrtle fasteneth and restoreth weary members and limbs, and therefore it tokeneth comforters of holy church.

Bartholomew (*Berthelet*), bk. xvii. § 101.

THE decoction of Myrtle made with wine withstandeth drunkenness if it be taken fasting.

Gerard's "Herbal," *s.v.*

IF a wayfaring man that hath a great journey for to go on foot, carry in his hand a stick or rod of the Myrtle-tree, he shall never be weary.

Holland's Pliny, bk. xv. ch. xxix.

Neat.

Neat's tongue.

MERCHANT OF VENICE, i. 1, 112.

NEAT'S tongue dried.

Middleton, "Blunt, Master Constable," i. 2, 188.

[Neat's tongue was boiled with red wine, stuffed with cloves and sugar, and served with red wine and prunes boiled together, and mustard (*Dawson*, "The Good Huswife's Jewel").]

Neat's foot.

TAMING OF THE SHREW, iv. 3, 17.

[Neat's foot was fricasseed, and the sauce was barberries or grape ("Good Huswife's Jewel").]

V. Ox.

Nettle.

i. KING HENRY IV., ii. 3, 10.

IT is a remedy against the venomous qualities of hemlock, mushrooms and quicksilver, and a counterpoison for henbane, serpents and scorpions. The oil of it takes away the stinging which the Nettle itself maketh.

Gerard's "Herbal," *s.v.*

CAST the water of any sick person newly made at night on red Nettles, and if the Nettles be withered and dead in the morning after, then the sick party is like to die of that disease, if they be green still, then he is like to live.

Lupton, "Notable Things," bk. iv. § 71.

THE virtue of Nettles is to force a woman that waters them to be as peevish for a whole day, and as waspish, as if she had been stung in the brow with a hornet.

Greene, "Quip for an Upstart Courtier."

Now are they plagued in purgatory, and he whips them with Nettles. *Tarleton*, "News out of Purgatory."

HIS hate to woman made Eupolis eat Nettle-pottage.

"Lady Alimony," i. 2.

HE who holds this herb in his hand with yarrow is secure from every fear and from every phantasy. And if it be put with the juice of house-leek, and the hand be anointed therewith, and the residue be put in water where are fishes, they will collect about his hand, and also about his net. And if the hand be taken out, forthwith they return to their own places where they were before.

Albertus Magnus, "Of the Virtues of Herbs."

Newt.

MIDSUMMER NIGHT'S DREAM, iii. 2, 11.

THIS is a little black Lizard of the water; the poison hereof is like the poison of vipers. This serpent is bred in fat waters and soils, and sometimes in the ruins of old walls. There is nothing in nature that so much offendeth it as salt. Being moved to anger, it standeth upon the hinder legs, and looketh directly in the face of him that hath stirred it, and so continueth till all the body be white, through a kind of white humour or poison, that it swelleth outward, to harm (if it were possible) the person that did provoke it. And by this is their venomous nature observed to be like the salamander, although their continual abode in the water maketh their poison the more weak. There be some Apothecaries which do use this Newt instead of skinks or crocodiles of the earth, but they are deceived in the virtues and operation, and do also deceive other, for there is not in it any such wholesome properties, and therefore not to be applied without singular danger.

Topsell, "History of Serpents," pp. 744.

EWTS' eggs be like to serpents' eggs, but they be less in quantity, and more glimy [gluey]; and be venomous, but they be less venomous than serpents' eggs.

Bartholomew (*Berthelet*), bk. xix. § 101.

V. Lizard.

Night-crow [=Night-heron].

iii. KING HENRY VI., v. 6, 45.

THE Night-crow loveth the night, and fleeth and seeketh his meat by night, and crieth in seeking, and their cry is hateful and odious to other birds. And they eat the eggs of doves and choughs, and fight with them. Also this bird hight Noctua [*i.e.*, the owl ?]; by night she may see, and when shining of the sun cometh, her sight is dim. The Island of Crete hath not this bird; if he cometh thither out of other lands, he dieth anon.

Bartholomew (*Berthelet*), bk. xii. § 27.

THIS kind of owl is dog-footed, and covered with hair; his eyes are like the glistering ice; against death he useth a strange whoop. *Batman's* addition to *Bartholomew, ut supra.*

BY night (as the vulgar think) the Night-crow seemeth with its hateful cry to portend the death of men. It is pleased with the human voice. The Night-crow is an antidote to bees, wasps, hornets and leeches. Its eggs given in wine for three days cause loathing of wine in drunkards. [This is a quotation from *Pliny*, bk. xxx., where Holland translates "owls' eggs."] *Hortus Sanitatis*, bk. iii. § 84.

Nightingale.

[In Philemon Holland's translation of Pliny is a most elequent description of the Nightingale, too long to quote (bk. x. ch. xxix.). Also (ch. xlii.) he says that Germanicus and Drusus had two Nightingales that were taught to speak Latin and Greek—yea, and were able to continue a long speech and discourse.]

Night-raven.

MUCH ADO ABOUT NOTHING, ii. 3, 84.

V. **Night-crow** and **Raven.**

THERE is another kind of Night-raven, black, of the bigness of a dove, flat-headed, out of the which groweth three long feathers, like the cop of a lap-wing, his bill grey, using a sharp voice; whose unaccustomed appearance betokeneth mortality. He preyeth on mice, weasels, and such like. *Batman's* addition to *Bartholomew*, bk. xii. § 27.

Nit.

LOVE'S LABOUR'S LOST, iv. 1, 150.

THESE are little, white, living creatures. The Philosopher affirms that they are called the eggs of lice. They are like to the flowers of Jessamine that grows with us. For as Jessamine brings flowers without seed, so Lice bring forth eggs without young ones in them.

Thos. Mouffet, "Theatre of Insects," bk. ii. ch. xxxv.

Nut.

AS YOU LIKE IT, iii. 2, 115.

DROPPING of the leaves thereof grieveth and noyeth other trees about, that be nigh thereto. The fruit thereof hath so great virtue, that if it be put among frog-stools and venomous meats, it spoileth and destroyeth and quencheth all the venom that is therein. And all manner apples that be closed in an hard skin, rind, or shell be called Nuts, as pines, chestnuts and filberts, and other such. The shadow of the Nut-tree grieveth them that sleep thereunder, and breedeth diverse sicknesses and evils, but the fruit thereof dyeth and cleanseth hair, and letteth the falling thereof. In great French Nuts [*i.e.*, walnuts or barnuts] generally the shape of the cross is printed therein.

Bartholomew (*Berthelet*), bk. xvii. § 108.

V. Chestnut, Filbert, Walnut, etc.

Nutmeg.

LOVE'S LABOUR'S LOST, v. 2.
WINTER'S TALE, iv. 3, 20.

THE more heavy the Nutmeg is in weight, and the more sweet in smell and sharp in savour, the better it is. The Nutmeg holden to the nose comforteth the brain and the spiritual members. *Bartholomew* (*Berthelet*), bk. xvii. § 109.

THE Nutmeg is good against freckles of the face [and] quickeneth the sight. There is not any so simple but knoweth that the heaviest, fattest and fullest of juice are the best, which may easily be determined by pricking the same with a pin or such like. *Gerard's* "Herbal," *s.v.*

As easily deciphered as the characters in a Nutmeg.

Lilly, "Midas," iv. 3.

Oak.

THE Oak is a tree that beareth mast, and is a fast tree and a sad, and dureth long time, with hard rind, and little pith or none, and there breedeth on the leaves a manner thing sour and unsavoury, and physicians call it gall.

Bartholomew (*Berthelet*), bk. xvii. § 134.

THE Oak is a tree with many boughs and branches, and, by reason of many fair leaves and broad, it causeth pleasant shadow, and beareth great plenty of fruit and of mast. The tree is durable and strong, and nigh unable to root; for stocks thereof laid under water turneth, as it were, into hardness of stone; and the longer time they be in such moist places, the more hard they be. And so for hard and durable matter and kind of such tree, misbelieved men made thereof images, and maumets [Mahomets] of false gods. *Bartholomew* (*Berthelet*), bk. xvii. § 84.

THE Oak-apples, being broken in sunder about the time of their withering, do foreshew the sequel of the year, as the expert Kentish husbandmen have observed, by the living things found in them; as if they find an ant, they foretell plenty of grain to ensue—if a spider, then (say they) we shall have a pestilence or some such like sickness to follow amongst men—if a white worm like a gentle or maggot, then they prognosticate murrain of beasts and cattle. These things the learned also have observed and noted.

Gerard's "Herbal," *s.v.*

[So Lupton, bk. iii. § 7: "If [the little worm in the oak-apple] doth fly away, it signifies wars; if it creep, it betokens scarceness of corn; if it turn about, then it foreshews the plague. This is the countryman's astrology, which they have long observed for truth."]

V. Gall.

Oats.

TEMPEST, iv. 1, 61.

OATS are used in many countries to make sundry sorts of bread, as in Lancashire, where it is their chiefest bread-corn for Jannocks [Oat-cakes], Haver-cakes, Tharf-cakes [Oat-cakes, unleavened], and those which are called generally Oaten-cakes; and for the most part they call the grain Haver, whereof they do likewise make drink for want of barley. Oatmeal is good for to make a fair and well-coloured maid to look like a cake of tallow, especially if she take next her stomach a good draught of strong vinegar after it. *Gerard's* "Herbal," *s.v.*

Oil.

OIL is the juice of herbs of olive, and the more fresh it is, the more noble it is, and the more slyly it cometh out of the hulls, the better it is and the more noble. If a man be under water with Oil in his mouth and spouteth out that Oil there in the water, all that is in the bottom, and hid by the ground, is the more clear and the more clearly seen of him. Kind of Oil maketh good savour in meat, and nourisheth light, and easeth, refresheth and comforteth weary bodies and limbs. Many diverse Oils be pressed out of many diverse things: as Oil of Olive, Oil of Nuts, Oil of Poppy, Oil of Almonds, of Raphans [*i.e.*, Radishes], Oil of Linseed, Oil of Hemp, and of other such. And Oil slayeth bees, and footless beasts with long and pliant bodies, if it be shed upon them, and vinegar turneth them again to life, if it be shed upon them.

Bartholomew (*Berthelet*), bk. xvii. § 112.

Olive.

AS YOU LIKE IT, iii. 5, 74.

THE tree thereof is most sad and fast, and pure and clean without rotting. And the Olive will not be hard beaten with stones and poles to gather the fruit thereof, as some men do that be unready and unwise, for it beareth the worse if it be so beaten.

Bartholomew (*Berthelet*), bk. xvii. § 111.

THE eagle is never stricken with thunder, nor the Olive with lightning. *Lilly*, "Sappho and Phaon," iii. 3.

SCORE a gallon of sack, and a pint of Olives to the Unicorn. *Beaumont and Fletcher*, "The Captain," iv. 2, 1.

Onion.

ALL'S WELL THAT ENDS WELL, v. 3, 321.
MIDSUMMER NIGHT'S DREAM, iv. 2, 43.

IT bringeth out venom, and quencheth biting of a wood hound, and helpeth in other venoms by bitings, and clarifieth the skin, and openeth pores. To eat too much of them breedeth madness and woodness, and maketh dreadful dreams, and namely if men that be new recovered of sickness eat thereof. *Bartholomew* (*Berthelet*), bk. xvii. § 42.

THE juice anointed upon a pilled or bald head in the sun, bringeth again the hair very speedily. The Onion being eaten, yea, though it be boiled, causeth head-ache, hurteth the eyes, maketh a man dim-sighted, dulleth the senses, and provoketh overmuch sleep, especially being eaten raw.

Gerard's "Herbal," *s.v.*

'Tis better than an Onion to a green wound i' the left hand made by fire, it takes out scar and all.

Webster, "Cure for a Cuckold," iv. 1.

Aye, Aye, Sir Lionel, they are my Onions; I thought to have had them roasted this morning for my cold. Gervase, you have not wept to-day; pray take your Onions.

John Cook, "Greene's Tu Quoque."

St. Thomas's Onions shall be sold by the rope at Billingsgate by the Statute.

"Pennyless Parliament of Thread-bare Poets," § 36 (1608).

Opal.

Twelfth Night, ii. 4, 77.

Opal is a stone distingued with colours of divers precious stones; therein is the fiery colour of the carbuncle, the shining purple of the amethyst, the bright green colour of emerald; and all the colours shine with a manner diversity. This stone breedeth only in Ind, and is deemed to have as many virtues as colours. This stone keepeth and saveth his eyes, that him beareth, clear and sharp and without grief;—and dimmeth other men's eyes that be about with a manner cloud, and smiteth them with a manner blindness, so that they may not see, neither take heed, what is done tofore their eyes. Therefore it is said, that it is most sure patron of thieves—[safest stone for thieves—*Bartholomew*].

Bartholomew (*Berthelet*), bk. xvi. § 73.

Orange.

Civil count, civil as an orange, and something of that jealous complexion.

Much Ado About Nothing, ii. 1, 304.

Orange-wife.

Coriolanus, ii. 178.

["Civil" is probably a pun upon the word "Seville."]

THE women [of Portugal] are for the most part like their Oranges, the fairer the outside the rottener within, and the sounder at the heart, the rougher the skin.

Heywood, "Challenge for Beauty," ii. 1.

[RECIPE] How to dress Oranges.

"The Widow's Treasure (1595).

[RECIPE] To confect Orange pills.

Second part of the "Good Huswife's Jewel," p. 42 (1597).

Two lemons and an Orange pill.

Bacchus' "Bounty" (1593).

HERE's New-Year's-Gift has an Orange and rosemary, but not a clove to stick in't.

Ben Jonson, "Christmas Masque" (1616).

WINE will be pleasant in taste and in savour and colour; it will much please thee, if an Orange or a lemon (stuck round about with cloves) be hanged within the vessel, that it touch not the wine. And so the wine will be preserved from fustiness and evil savour.

Lupton, "Notable Things," bk. ii. § 40.

[An Orangeado-pie is mentioned as a delicacy by Dekker in one of his plays.
Neither Pliny nor Bartholomew mentions oranges.]

Osier.

LOVE'S LABOUR'S LOST, iv. 2, 112.

V. Willow.

WHAT is this snare to which young virgins haste,
But like the Osier wheel in rivers placed?
The fish yet free to enter wind about,
Whilst they within are labouring to get out.

Heywood, "Anna and Phyllis," embl. 2.

Osprey.

As is the osprey to the fish, who takes it
By sovereignty of nature.

CORIOLANUS, iv. 7, 34.

THE Osprey only, before her little ones be feathered, will beat and strike them with her wings, and thereby force them to look full against the sun-beams; now if she see any one of them to wink, or their eyes to water at the rays of the sun, she turns it with the head forward out of the nest, as a bastard, and not right, nor none of hers,—but bringeth up and cherisheth that whose eye will abide the light of the sun, as she looks directly upon him. Moreover these Ospreys are not thought to be a several kind of Eagles by themselves, but to be mongrels, and engendered of diverse sorts. When [eagles] have cast [their young] off, the Ospreys, which are near of kin unto them, are ready to take them, and bring them up with their own birds. *Holland's Pliny*, bk. x. ch. iii.

OSPREYS are called *ossifragi* [or bone-breakers], because they drop bones from on high, and break them.

Hortus Sanitatis, bk. iii. § 89.

THE Osprey oft here seen, though seldom here it breeds,
Which over them the fish no sooner doth espy,
But, betwixt him and them by an antipathy,
Turning their bellies up, as though their death they saw,
They at his pleasure lie, to stuff his gluttonous maw.

Drayton, "Polyolbion," song xxv. (quoted by Steevens).

Ostrich.

ii. KING HENRY VI., iv. 10, 31.

THE Ostrich hath a body as a beast, and feathers as a fowl, and also he hath two feet and a bill as a fowl. And for he is somedeal shaped as a bird, he hath many feathers in the nether part of the body, and hath two feet as a fowl, and is cloven-footed as a four-footed beast; and is so hot, that he swalloweth, and defieth [digesteth] and wasteth iron. And when the time is come that they shall lay eggs, they heave up their eyes, and behold the stars that hight Pleiades, for they lay no eggs, but when that constellation ariseth and is seen. And about the month of June, when they see those stars, they dig in gravel, and lay there their eggs, and cover and hide them with sand; and when they have left them there, they forget anon where they have laid them, and come never again thereto; but the gravel is chauffed [warmed] with the heat of the sun, and heateth the eggs that be hid, and breedeth birds therein, and bringeth them forth; and when the shell is broken, and birds come out, then first the mother gathereth and nourisheth them,—and the bird that she despised in the egg, she knoweth when it is come out of the egg. Also the Ostrich hateth the horse by kind, and is so contrary to the horse, that he may not see the horse without fear. And if an horse come against him, he raiseth up his wings as it were against his enemy, and compelleth the horse to flee with beating of his wings.

Bartholomew (*Berthelet*), bk. xii. § 33.

Cloven hoofs they have like red deer, and with them they fight, for good they be to catch up stones withal, and with their legs they whirl them back as they run away against those that chase them. But the veriest fools they be of all others;—for as high as the rest of their body is, yet if they thrust their head and neck once into any shrub or bush, and get it hidden, they think then they are safe enough, and that no man seeth them.

Holland's Pliny, bk. x. ch. i.

SOME reasonless creatures likewise are by nature bold, as Ostriches. *Holland's Pliny*, bk. xi. ch. xxxvii.

THE Ostrich has a small bone under its wings, by which it purges itself in the side, and shakes it when it is provoked to anger. It has a very strong skin, by which with its feathers it is protected from the troublesome cold.

Hortus Sanitatis, bk. iii. § 109-10.

Sir Gosling:—Sing or howl, or I'll break your Ostrich egg-shell there.

Birdlime:—My egg hurts not you.

[Birdlime is an elderly lady with not the best of characters.] *Webster*, "Westward Ho!" v. 3.

[OSTRICH] a foolish bird that forgetteth his nest, and leaveth his eggs for the sun and sand to hatch, that eateth any thing, even the hardest iron, that heareth nothing.

Purchas' "Pilgrims," p. 560 (ed. 1616).

Otter.

Neither fish nor flesh.

i. KING HENRY IV., iii. 3, 142-4.

THERE is no doubt but this beast is of the kind of beavers, saving in their tail, for the tail of a beaver is fish, but the tail of an Otter is flesh. It hath very sharp teeth, and is a very biting beast. So great is the sagacity and sense of smelling in this beast, that he can directly wind the fishes in the water a mile or two off. There is a kind of Assa called Benjoin, a strong herb, which, being hung in a linen cloth near fish-ponds, driveth away all Otters and beavers. The skin doth not lose its beauty by age, and no rain can hurt it, and is sold for seven or eight shillings; thereof they make fringes in hems of garments, and face about the collars of men and women's garments, and the skin of the Otter is far more precious than the skin of the beaver. *Topsell*, "Four-footed Beasts," *s.v.*

I MARVEL how it came into the writer's head to affirm that the beaver constraineth the Otter in the winter-time to trouble the water about her tail to the intent it may not freeze. *Ibid.*

Ounce.

MIDSUMMER NIGHT'S DREAM, ii. 3.

SOME have said when a man or beast is bitten with an Ounce, presently mice flock unto him, and poison him with their urine. The gall of this beast is deadly poison; it hateth all creatures, and destroyeth them.

Topsell, "Four-footed Beasts," *s.v.*

THE Ounce does not eat its prey, until it has hung it up on high, but when it comes to a tree, it carries its prey to the topmost branch, and eats it hanging. [Then follows the above curious statement about the Ounce-bite and the mice, with a story of a man bitten by an Ounce, "who had himself carried out to sea in a bark," and so baffled the mice.]

Hortus Sanitatis, bk. ii. § 158.

Ousel.

MIDSUMMER NIGHT'S DREAM, iii. 1, 128.

THE Ousel or blackbird is white in Achaia. The Ousel purges disgust to meat annually with laurel-leaves. The Ousel changes its colour from black to russet, sings in the summer, stutters in winter, changes about the solstice its bill, which is transformed into ivory in year-old cocks. The tame Ousel eats flesh against nature. The Ousel like other birds does not shed its plumage, but changes its bill to a white colour every year. And in the winter for fatness it can scarcely fly.

Ibid., bk. iii. § 74.

IF the feathers of the right wing of an Ousel be hung up on a red thread, which has never been used, in the middle of a house, no one will be able to sleep in that house, until the wing has been taken down. And if its heart be put under the head of a sleeper, and he be questioned, he will tell with a loud voice all that he has done. And again if it be put in well-water with the blood of a hoopoo, and mixed together, and then rubbed on the temples of any man, he grows weak even to death.

Albertus Magnus, "Of the Virtues of Animals."

Owl.

The Owl is a wild bird charged with feathers, but she is always with-holden with sloth, and is feeble to fly, and dwelleth by graves by day and by night, and in chincs. And diviners tell that they betoken evil; for if the Owl be seen in a city, it signifieth destruction and waste. The chough fighteth with the Owl, and taketh the Owl's eggs, and eateth them by day, and the Owl eateth the chough's eggs by night. The crying of the Owl by night tokeneth death. The Owl is fed with dirt, and with other unclean things. *Bartholomew* (*Berthelet*), bk. xii. § 5.

If the heart of an Owl be laid on the left side of a sleeping woman, she will tell all that she hath done. The feet of Owls burnt with the herb plantain help against serpents. They put the ashes of Owls' eyes on madmen.
Hortus Sanitatis, bk. iii. § 16.

The Howlet, Screech-owl, etc., when they be hatched come forth of their shells with their tail first; and by reason of their heads so heavy, the eggs are turned with the wrong end downward. *Holland's Pliny*, bk. x. ch. xvi.

It is a pretty sight to see the wit and dexterity of these Howlets, when they fight with other birds; for when they are overlaid and beset with a multitude of them, they lie upon their backs and with their feet make shift to resist them. The falcon by a secret instinct and society of nature, seeing the poor Howlet thus distressed, cometh to succour and taketh equal part with him, and so endeth the fray. Howlets for sixty days in winter keep close and remain in covert, and they change their voice into nine tunes.
Ibid., ch. xvii.

For the sting of bees, wasps and hornets,—for the biting also of those horse-leeches called blood-suckers, the Howlet is counted a sovereign remedy, by a certain antipathy in nature. *Ibid.*, bk. xxix. ch. iv.

If any man put the heart of an Owl under his armpit, no dog will bark at him, but will keep silence; and if the heart and right foot with its breath [*cum anima sua*] be hung on a tree, the birds will gather together upon that tree. *Albertus Magnus*, "Of the Virtues of Animals."

Ox.

THE dewlap or freshlap that hangeth down under his throat and stretcheth to the legs is a token of gentleness and nobility in an Ox. Oxen wax fat by washing with hot water. And Oxen with straight horns be accounted excellent in work, and black Oxen with little horns be accounted less profitable to working. Of Ox-horns be made tapping and nocks to bows, to arbalisters, and arrows to shoot against enemies, and breast-plates and other armour by the which unstrong places of man's body be warded and defended against shot and smiting of enemies. And of Ox-horns be lanterns made to put off darkness, and combs to right and to cleanse heads of filth. Also writers and painters use the horns, and keep in them divers colours at best. Also there is a little beast like to *scarabæus*, and hight *Burestis*, and this burestis beguileth and betrayeth the Ox in the grass, and that is for the Ox treadeth on him. For this burestis lieth among herbs and grass that the Ox loveth, and hideth him therein, and the Ox swalloweth this beast burestis, and he chafeth suddenly the liver of the Ox, and maketh him break with great pain and sorrow.

Bartholomew (*Berthelet*), bk. xviii. § 13.

THEOPHRASTUS approves of feeding Oxen with fish, but only with live fish. *Hortus Sanitatis*, bk. ii. § 14.

THERE are Oxen in India which will eat flesh like wolves, and have but one horn and whole hoofs; some also have three horns. There be Oxen in Leuctria have their ears and horns growing both together forth of one stem. The Oxen of the Geramants and all other neat among them feed with their necks doubled backward, for by reason of their long and hanging horns, they cannot eat their meat holding their heads directly straight. There be Oxen in Phrygia which are of a flaming red colour, of a very high and winding neck; their horns are not like any other in the world, for they are moved with their ears, turning in a flexible manner sometime one way and sometime another. In some countries they wash them all over with wine for two or three days together, which doth wonderfully tame them, though they have been never so wild. If a wild Ox be tied with a halter of wool, he will presently wax tame. If the Ox bend to the right side and lick that, it presageth a storm; but if he bend to the left side, he foretelleth a calm, fair day. In like manner, when he loweth and smelleth to the earth, or when he feedeth fuller than ordinary, it betokeneth change of weather. If a wolf's tail be hanged in the rack or manger where an Ox feedeth, he will abstain from eating. If seed be cast into the earth out of an Ox's horn, it will never spring up well out of the earth, or at the least not so well as when it is sowed by the hand of man. Of the teeth of Oxen I know no other use but scraping and making paper smooth with them; their gall being sprinkled among seed which is to be sown maketh it come up quickly, and killeth field-mice that taste of it. The dung of Oxen is beneficial to bees if the hives be anointed therewith, for it killeth spiders, gnats and drone-bees. When a man biteth any other living creature, seethe the flesh of an Ox or a calf, and after five days lay it to the sore, and it shall work the ease thereof. If one make a small candle of paper and cow's marrow, setting the same on fire, under his brows or eye-lids which are bald without hair, and often anointing the place, he shall have very decent and comely hair grow thereupon. There is in the head of an Ox a certain little stone, which only in the fear of death he casteth out at his mouth; if this stone be taken from them suddenly by cutting the head, it doth make children to breed teeth easily, being soon tied about them. When the bee hath tasted of the flower of the corn-tree, she presently dieth, except she taste the urine of a man or an Ox. *Topsell*, "Four-footed Beasts," *s.v.*

V. Cow and Bull.

Oxlip.

WINTER'S TALE, iv. 4, 125.

OXLIPS, so called because oxen and cows delight in eating them. *Minsheu's* Dictionary, *s.v.*

OXLIPS or Paigles. *Gerard's* "Herbal," *s.v.*

Oyster.

As You Like It, v. 4, 64.

Fish conceive of dew only without peasen [*i.e.*, spawn] and without milk [or milt] as Oysters and other shell-fish. When the moon faileth such fishes be void; and the waxing of the moon increaseth the humour, and the humour vanisheth, when the moon vanisheth. By night shell-fish come to cliffs, and conceive pearls of the dew of heaven.

Bartholomew (*Berthelet*), bk. xiii. § 29.

The ashes of an Oyster-shell calcined and incorporate with honey rid away wrinkles, and make women's skin to lie smooth and even.

Holland's Pliny, bk. xxxii. ch. vi.

The gaping Oyster, entertaining stones
By th' crab injected, is despoiled at once.

Heywood, "Anna and Phyllis," embl. 29.

[Colchester Oysters were renowned in Shakespeare's time (*Massinger*, "New Way to pay Old Debts," iv. 1, and *Nash's* "Lenten Stuff). So Kentish Oysters (*Greene's* "Tu quoque"), Wallfleet Oysters ("Vox Graculi," quoted by *Brand*, vol. i. p. 15). Red wine was drunk with Oysters ("Pennyless Parliament of Threadbare Poets," § 3, and *Lilly*, "Mother Bombie," ii. 5: "He that had a cup of red wine to his Oysters was hoisted in the Queen's subsidy-book"). They were seasoned with vinegar (*Middleton*, "Spanish Gipsy," iv. 1, 10), or eaten pickled (*Greene's* "Tu quoque"), or in pies ("Lingua," v. 8, and *Ben Jonson*, "Cynthia's Revels," ii. 2), of which the ingredients, according to an eighteenth-century recipe in *Nares'* Glossary, were pepper, nutmeg, salt, currants, dates, barberries, mace, butter, lemons, anchovies, white wine, sugar, and the juice of an orange. George Peele, who was always distressed for money, was found all alone at a *peck* of Oysters (*Peele's* "Merry Conceited Jests"). "You may open an Oyster or two before grace" (*Beaumont and Fletcher*, "Maid in the Mill," iv. 1), and "capons crammed full of itching Oysters" (*ibid.*, "Women Pleased," ii. 4), need no explanation beyond the context.]

At Venice the Oysters are very dear, some twenty for a lira [*i.e.*, nine-pence].

Fynes Moryson, "Itinerary," part iii. p. 112.

V. Crab.

Paddock.

HAMLET, iii. 4, 190.
MACBETH, i. 1, 9.

IT is apparent that there be three kinds of frogs of the earth:—the first is the little green frog; the second is this Paddock, having a crook back; and the third is the toad. This second kind is found deep in the earth, in the midst of rocks and stones. Such as these are found near Tours in France, among a red sandy stone, whereof they make the mill-stones, and therefore they break that stone all in pieces before they make the mill-stone up, lest while the Paddock is included in the middle, and the mill-stone going in the mill, the heat should make the Paddock swell, and so the mill-stone breaking, the corn should be poisoned.

Topsell, "History of Serpents," p. 725.

V. Toad, Frog.

Palm-tree.

AS YOU LIKE IT, iii. 2, 186.

IF the male Palm be felled, then is the female barren after two days out. The more noble and old the Palm is the better the fruit thereof. And the Palm-tree beareth no fruit tofore an hundred years, and then it hath the first perfect and complete virtue.

Bartholomew (*Berthelet*), bk. xvii. § 116.

Pansy.

Pansies that's for thoughts.
HAMLET, iv. 5, 175.

HEART'S-EASE is named in English: Heart's ease, Pansies, Live in idleness, Cull me to you, and Three faces in a hood. *Gerard's* "Herbal," *s.v.*

V. Heart's-ease, Violet.

Panther.

TITUS ANDRONICUS, i. 1, 493.

PANTHER is friend to all beasts save the dragon, for him he hateth full sore. And is a beast painted with small round speckles, so that all his skin without seemeth full of eyes. And this beast whelpeth but once, and the cause thereof is openly known:—for when the whelps wax strong in the mother's womb, they hate the mother, and rend her womb with claws, and therefore the mother letteth pass and whelpeth them. The Panther hateth the dragon, and the dragon fleeth him. And when he hath eaten enough at full, he hideth him in his den, and sleepeth continually nigh three days, and riseth after three days and crieth, and out of his mouth cometh right good air and savour, and is passing measure sweet; and for the sweetness all beasts follow him; and only the dragon is afeard when he heareth

his voice, and fleeth into a den, and may not suffer the smell thereof, and faileth in himself, and loseth his comfort, for he weeneth that his smell is very venom. And all four-footed beasts have liking to behold the diverse colours of the Panthers and tigers, but they be afeard of the horribleness of their heads; and therefore they hide their heads, and toll [*i.e.*, draw] the beasts to them with fairness of that other deal of the body, and take them when they come so tolled, and eat them. And though he be a right cruel beast, yet he is not unkind to them that help and succour him in any wise.

Bartholomew (*Berthelet*), bk. xviii. § 82.

THEY have one mark on their shoulder resembling the moon, growing and decreasing as she doth, some times showing a full compass, and otherwhiles hollowed and pointed with tips like horns.

Holland's Pliny, bk. viii. ch. xvii.

A BROTH made of such pullein [cocks and capons] hath a singular virtue—for neither lions nor Panthers will set upon those persons who are bathed with their decoction, especially if there were any garlic sodden therein.

Ibid., bk xxix. ch. iv.

THE Panther so
Breathes odours precious as the Sannatic gums
Of Eastern groves, but the delicious scent,
Not taken in at distance, chokes the sense
With the too musky flavour.

Glapthorne, "Hollander."

THE Panther though his skin be fair, yet his breath is infectious. *Reynolds*, "God's Revenge against Murder," p. 257.

V. **Leopard** and **Pard.**

Paraquito.

i. KING HENRY IV., ii. 388.

V. **Parrot.**

Pard.

TEMPEST, iv. 1, 262.

THE Pard is the most swift beast, with many diverse colours and round specks as the panther, and reseth to blood, and dyeth in leaping, and varieth not from the panther, but the panther hath more white specks. The Pard when he is sick eateth man's dirt because of medicine; hunters hang that dirt on a tree, and he goeth up to it; and the hunters slay him. And is lecherous, and gendereth with the lioness:—of that bastard generation cometh the leopard [*v.* **Lioness**]. The Pard is cruel when his whelps be stolen.

Bartholomew (*Berthelet*), bk. xviii. § 83.

Of the Panther, commonly called a Pardal, a Leopard, and a Libbard. There have been so many names devised for this one beast that it is grown a difficult thing to define it perfectly. The panther is the female, and the Pard the male. When the lion covereth the Pardal, then is the whelp called a leopard or libbard, but when the Pardal covereth the lioness, then it is called a panther. The only difference betwixt the leopard, Pardal and lion is that the leopard or Pardal have no manes. The greatest they call panthers, the second they call Pardals, and the third, least of all, they call leopards, which in England is called a cat of the mountain. And truly in my opinion they are all one kind of beast, and differ in quantity only through adulterous generation. The leopard is a wrathful and an angry beast, and, whensoever it is sick, it thirsteth after the blood of a wild cat, and recovereth by sucking that blood. Above all other things, it delighteth in the camphor-tree, and therefore lieth underneath it to keep it from spoil; and in like sort the panther delighteth in sweet gums and spices, and therefore no marvel if they cannot abide garlic, because it annoyeth their sense of smelling; and if the walls of one's house or sheep-cote be anointed with the juice of garlic, both panthers and leopards will run away from it. The leopard is sometimes tamed and used for hunting; yet such is the nature of this beast, as also of the Pardal, that if he do not take his prey at the fourth or fifth jump, he destroyeth whomsoever he meeteth, yea, many times his hunter. Therefore the hunters have always a regard to carry with them a lamb or a kid, wherewithal they pacify him after he hath missed his game. The panthers of Lycia and Caria are very long, but yet weak and without carriage, being not able to leap far, yet is their skin so hard as no iron can pierce. There is a beast called *Bitis* not unlike to the vulgar leopards in all parts, except that it wanteth a tail; and if this beast be seen by a woman, it will instantly make her to be sick. Great is the love of the panther to all spices and aromatical trees. The female panther is more generous than the male. There is great hatred and enmity between the hyæna and the panther, for in the presence of the hyæna, the Pardal dare not resist; and if there be a piece of an hyæna's skin about either man or beast, the panther will never touch it; and if their skins after they be dead be hung up in the presence of one another, the hair will fall off from the panther. If anything be anointed with broth wherein a cock hath been sodden, neither panthers nor lions will ever touch it. Leopards are afraid of a certain tree called Leopard's-tree. Panthers are also afraid of the skull of a dead man, and run from the sight thereof. Likewise in Armenia there are certain fishes which are poison to lions, bears, wolves, lynxes and panthers; the powder of this fish the inhabitants put into the sides and flesh of their sheep, goats and kids without all harm to these beasts,—but if the panthers or any ravening beast come and devour any of those sheep so dressed, presently they die by poison. In hunting of wild beasts the wary woodman must make good choice of his horse, not only for the mettle and agility—which are very necessary,—but also for the colour; for the grey horse is fittest for the bear and most terrible to him, the yellow or fire-colour against the boar, but the brown and reddish-colour against the panther. Leopards and panthers also love wine above all other drink. If the skin or hide of a leopard being taken and flayed be covered or laid upon the ground, there is such force and virtue in the same that any venomous or poisonous serpent dare not approach into the same place where it is so laid. The gall of a panther being received into the body either in meat or drink, doth instantly and out of hand kill or poison him which doth so receive it.

Topsell, "Four-footed Beasts," *s.v. Pardal.*

Parrot.

MERCHANT OF VENICE, iii. 5, 51

THE Parrot can endure any other kind of water in any wise, but dies of rain; and therefore they build in Mt. Gilboa, because there it seldom or never rains. It drinks wine freely, and is much pleased with the sight of a virgin.

Hortus Sanitatis, bk. iii. § 102.

SHE hath an head as hard as is her beak; when she learns to speak, she must be beaten about the head with a rod of iron, for otherwise she careth for no blows.

Holland's Pliny, bk. x. ch. xlii.

Parsley.

Parsley to stuff a rabbit.
TAMING OF THE SHREW, iv. 4, 100.

[Parsley is one of the herbs used to make broth and farcing enumerated in "The History of Jacob and Esau," iv. 5.]

A SALAD of Parsley and the herb patience.
"Look About You," i. 10.

GOOD man-mender,
Stop me with some Parsley like stuffed beef,
And let me walk abroad.
Beaumont and Fletcher, "The Chances," iii. 2.

TOUGH Welsh Parsley which in our vulgar tongue is strong hempen halters. *Ibid.*, "The Elder Brother," i. 2.

Partridge.

ii. KING HENRY VI., iii. 2, 191.

THE Partridge is an unclean bird, for strong liking of lechery forgetteth the sex and distinction of male and female. And is so guileful that the one stealeth the eggs of the other, and sitteth abrood on them; but this fraud hath no fruit, for when the birds be haught [grown], and hear the voice of their own mother, they forsake her that brooded them when they were eggs, and kept them as her own birds, and turn and follow their own mother natural. And the Partridge travaileth not in laying and in brooding, like as other fowls do. And at the noise of a little bell, he fleeth about upon the ground, and falleth into the gin or net ere he be ware. The Partridge's gall, with even weight of honey, cleareth much the sight; and therefore it shall be kept in a silver box.
Bartholomew (*Berthelet*), bk. xii. § 30.

A PARTRIDGE will cry aloud, and will tear or break the cage or coop where she is fed, if there be any deadly medicine or poison prepared within the same house, which she doth feel presently, through a wonderful special and rare gift of nature. *Lupton*, "Notable Things," bk. x. § 99.

THE Partridge, though cunning in many things, is foolish in this, that where she can hide her head, she believes that her whole body is hidden, and when she sees no one, she thinks that no one sees her. *Hortus Sanitatis*, bk. iii. § 98.

PARTRIDGES do so fortify and impale their nests with thorns and twigs of shrubs and bushes, that they be sufficiently fenced against the invasion of wild beasts. Partridges live 16 years. *Holland's Pliny*, bk. x. ch. xxxiii.

IN Paphlagonia Partridges have two hearts.

Ibid., bk. xi. ch. xxxvii.

SOME creatures there be that will never be fat, as the hare and Partridge. *Ibid.*

Pea.

LOVE'S LABOUR'S LOST, v. 2, 316.

THE great Pease is called Roman Pease, or Garden Pease, of some, Branch-Pease, French Pease, and Rouncivals.

Gerard's "Herbal," *s.v.*

[Rouncivals, Tusser tells us (October's "Husbandry," st. 30, and November, st. 7), were gray Pease. He mentions also Hastings (October, st. 32, margin, and November, st. 7) and Fulhams (October, *ut supra*). A dish of buttered Pease is mentioned in the Interlude of "The Trial for Treasure," and in *Greene's* "Quip for an Upstart Courtier."

White Pease-pottage consisted of boiled white Pease with butter and verjuice and a little fine powder of March, and was served upon sops (second part of the "Good Huswife's Jewel," p. 26), and permission is accorded to boil porpoise and seal in your Pease.

The price of early Pease is given by Middleton as ten groats the cod ("Blurt, Master Constable," iii. 3, 123), or four nobles a peck may be nearer the mark (*Dekker's* "Bachelor's Wedding."]

Peach.

[Shakespeare only uses "Peach-coloured."]

MEASURE FOR MEASURE, iv. 3, 12.
ii. KING HENRY IV., ii. 2, 19.

[To the four sorts described by Gerard (viz., the White, the Red, the *d'Avant*, and the Yellow), Johnson (1633) adds those "choice ones" to be had from his friend Mr. Millen in Old

Street, *i.e.*, Nutmeg (two sorts), Queen's, Newington, Grand Carnation, Black, Mellicotton, Roman, Alberza, Island, and Du Troy. Also of "that kind of Peach which some call Nectarine, the Roman Red (the best of fruits), the Bastard Red, the little dainty Green, the Yellow, the White, the Russet, which is not so good as the rest."

Evelyn ("Kalendarium Hortense") gives the following: Admirable, Alberge (Sir H. Capel's), Alberge (small yellow), Almond Violet, Bourdin, Belle Cheuvreuse, Elruge Nectarine (excellent), Maudlin, Mignon, Morella, Musk Violet, Murrey Nectarine, Nutmeg (white, red), Man Peach, Persique, Rambullion, Sion (excellent), Orleans, Savoy Mellicotton, etc.]

PEACHES, by making the belly slippery, cause other meats to slip down the sooner. The kernel within the Peach-stone stamped small, and boiled with vinegar until it be brought to the form of an ointment, is good to restore and bring again the hair. *Gerard's* "Herbal," *s.v.*

To confect Peaches after the Spanish fashion.

Second part of "Good Huswife's Jewel," p. 43.

Peacock.

i. KING HENRY VI., iii. 3, 6.

HIS flesh is so hard that unneath [with difficulty] it rotteth, and is full hard to seething. The Peacock is a bird that loveth not his children; for the male seeketh out the female, and seeketh out her eggs for to break them, and the female dreadeth that, and hideth busily her eggs, lest the Peacock might soon find them. And the Peacock hath foulest feet and rivelled [*i.e.*, wrinkled]. And he wondereth of the fairness of his feathers, and reareth them up, as it were a circle about his head, and then he looketh to his feet, and seeth the foulness of his feet, and, like as he were ashamed, he letteth his feathers fall suddenly, and all the tail downward, as though he took no heed of the fairness of his feathers. And hath a voice of a fiend, head of a serpent, pace of a thief.

Bartholomew (*Berthelet*), bk. xii. § 31.

BY his voice he frightens serpents, and drives away all venomous animals, so that they dare not stay where his voice is often heard. The Peacock when he ascends on high betokens rain. *Hortus Sanitatis*, bk. iii. § 93.

Pear.

ALL'S WELL THAT ENDS WELL, i. 1, 178.
MERRY WIVES OF WINDSOR, iv. 5, 100.

Poperin pear.
ROMEO AND JULIET, ii. 1, 38.

FEW Pears weigh heavier than many apples, if they be on a beast's back. And Pears have this property, that, if they be sod with toadstools, they take away from them all grief and malice. Powder or ashes of wild Pears drunken helpeth against fungus, toadstools. Alway after eating of Pears, wine shall be drunk, for without wine Pears be venom.

Bartholomew (*Berthelet*), bk. xvii. § 124.

[Gerard ("Herbal," *s.v.*) names the following Pears: the Catharine, the Jenneting, St. James, Pear Royal, the Bergamot, the Quince Pear, the Bishop's, and the Winter; also of wild Pears: the great and small Choke-Pears, the wild Hedge-Pear, the wild Crab-Pear, the lousy Wild-Pear, and the Crow-Pear. He states that all the garden Pears "before specified, and many sorts more, and those most rare and good, are growing in the ground of Master Richard Pointer, a most cunning and curious graffer and planter of all manner of rare fruits, dwelling in a small village near London, called Twickenham, and also in the ground of an excellent graffer and painful planter, Mr. Henry Banbury, of Tothill Street, near Westminster, and likewise in the ground of a diligent and most affectionate lover of plants, Mr. Warner, near Horsely Down, by London."

Malone says that the Poperin Pear came from Poperingue, near Ypres, in French Flanders.

Pear-pies are mentioned in *Preston's* "Tragedy of Cambyses."]

V. **Warden.**

Pearl.

AS YOU LIKE IT, v. 4, 63.

MARGUERITE (*i.e.*, Pearl) is chief of all white precious stones. It breedeth in flesh of shell-fish, and is sometime found in the brain of the fish; and is gendered of the dew of heaven, the which dew shellfish receive in certain times of the year. Of the which marguerites some be called Unions, and have a convenable name, for only one is found, and never two or more together. And those that be conceived of the morrow [*i.e.*, morning] dew be made dim with the air of the even-tide. And some be found kindly [*i.e.*, naturally] pierced, and those be better than other; and some be pierced by craft. And they have virtue comfortative, either of all the whole kind, or else, because they are besprung [sprinkled] with certain speciality, they comfort the limbs; for by constraining and coarcting, they cleanse them of superfluous humours. And the more of dew and air that is drawn in, the more and the greater they be, but no marguerite groweth passing of half a foot. Also if that lightning or thundering fall, when the marguerite should breed of the dew that is drawn in, the shell closeth by sudden fear, and so the gendering faileth, and is cast out. *Bartholomew* (*Berthelet*), bk. xvi. § 62.

THE shell that is the mother-of-pearl, as soon as it perceiveth and feeleth a man's hand within it, by and by she shutteth, for well wotteth she that for her riches she is sought for; but let the fisher look well to his fingers, for if she catch his hand between, off it goeth. Some say that these mother-pearls have their kings and captains as bees have. *Holland's Pliny*, bk. ix. ch. xxxv.

Pelican.

HAMLET, iv. 5, 146.

A PELICAN is a bird of Egypt. And there be two manner Pelicans: one dwelleth in water and eateth fish, and the other dwelleth on land and loveth wilderness, and eateth venomous beasts, as lizards and other such. When the Pelican's children be haught [grown], and begin to wax hoar, they smite the father and the mother in the face, wherefore the mother smiteth them again and slayeth them. And the third day the mother smiteth herself in her side, that the blood runneth out, and sheddeth that hot blood upon the bodies of her children, and by virtue of the blood the birds, that were before dead, quicken again. The Pelican is a bird with great wings, and most lean; for all that he swalloweth passeth forth anon behind; for he hath a right slipper gut, and therefore he may not hold meat till it be incorporate. And the serpent hateth kindly [*i.e.*, by nature] this bird; wherefore when the mother passeth out of the nest to get meat, the serpent climbeth on the tree, and stingeth and infecteth the birds; and when the mother cometh again, she maketh sorrow three days for her birds. Then she smiteth herself in the breast, and springeth blood upon them, and reareth them from death to life; and then for great bleeding the mother waxeth feeble, and the birds be compelled to pass out of the nest to get themselves meat. And some of them for kind love feed the mother that is feeble; and some be unkind, and care not for the mother, and the mother taketh good heed thereto, and when she cometh to her strength, she nourisheth and loveth those birds that fed her in her need, and putteth away her other birds as unworthy and unkind, and suffereth them not to dwell nor live with her.

Bartholomew (*Berthelet*), bk. xii. § 29.

THE Pelican lives on the milk of the crocodile, and therefore of choice follows the crocodile.

Hortus Sanitatis, bk. iii. § 97.

IF it be hung on the neck of any bird, it will fly continually until it falls dead; and its right foot after three months, from the moisture and warmth which the bird has, will be generated alive, and will move itself. [N.B. The statement is (perhaps intentionally) obscure.]

Albertus Magnus, "Of the Virtues of Animals."

Pepper.

TWELFTH NIGHT, iii. 4, 158.

PEPPER is the seed or the fruit of a tree that groweth in the south side of the hill Caucasus, and serpents keep the woods that Pepper groweth in; and when the woods of Pepper be ripe, men of that country setteth them on fire, and chacen (chase) away the serpents by violence of fire,—and by such burning the green of Pepper, that was white by kind, is made black and rivelly [*i.e.*, wrinkled]. And of Pepper be three manner kinds, long (and that is not ripe), white, and black. And black Pepper is most virtuous, and may longest be kept in heat, and is stronger than other Pepper—and the more heavy it is, the better it is, and the more new. And it is feigned new by fraud and guile of merchandise; for they cover the most eldest Pepper, and spring [*i.e.*, sprinkle] thereon ore of silver, or of lead, for it should so seem fresh and new because of the white husk. *Bartholomew* (*Berthelet*), bk. xvii. § 131.

[Pepper came also from Amboyna, in the East Indies (*Beaumont and Fletcher*, "Fair Maid of the Inn"), and from Guinea (*Webster*, "Devil's Law-Case.")]

Pheasant.

WINTER'S TALE, iv. 4, 769.

THE Pheasant is caught thus: sometimes the fowler, being covered with a cloth on which this bird is painted, shows himself to the Pheasant, which follows the man so covered, who does not retire nor fly, and at last, the Pheasant is caught in a net by the fowler's mate lying in wait. This bird is sad in rainy weather, and hides itself in thickets and woods. It digs its beak into the ground, and believes itself to be altogether hidden in this way. It moults from fatness. *Hortus Sanitatis*, bk. iii. § 46.

PHEASANTS will die of lice, unless they bestrew themselves with dust. *Holland's Pliny*, bk. xi. ch. xxxiii.

MEN may talk of country Christmasses, and court-gluttony,
Their thirty-pound buttered eggs, their pies of carps' tongues,
Their Pheasants drench'd with ambergris.

Massinger, "City Madam," i. ii. 3.

A PHEASANT, larded.

Massinger, "New Way to pay Old Debts," i. 2.

YOU shall eat nothing but shrimp-porridge for a fortnight; and now and then a Pheasant's egg supped with a peacock's feather. *Brome*, "The Sparagus Garden," ii. 3.

PARTRIDGES, Pheasants, woodcocks and the like, in some places so abound with us, as they bear little or no price.

Fynes Moryson, "Itinerary," part iii. p. 134.

Phœnix.

TEMPEST, iii. 3, 23.
iii. KING HENRY VI., i. 4, 36.

PHŒNIX is a bird, and there is but one of that kind in all the wide world; therefore lewd men wonder thereof. Phœnix is a bird without make [mate], and liveth three

hundred or five hundred years; when the which years be passed, she feeleth her own default and feebleness, and maketh a nest of right sweet smelling sticks, that be full dry, and in summer when the western wind bloweth, the sticks and the nest be set on fire with burning heat of the sun, and burneth strongly. Then this bird Phœnix cometh wilfully into the burning nest, and is there burnt to ashes, among these burning sticks. And within three days, a little worm is gendered of the ashes, and waxeth little and little, and taketh feathers, and is shaped and turned to a bird. And is the most fairest bird that is, most like to the peacock in feathers, and loveth wilderness, and gathereth his meat of clean greens and fruits.

Bartholomew (*Berthelet*), bk. xii. § 14.

THE Phœnix is born among the Arabs, a bird of the size of the eagle, its head adorned with plumes upstanding, its jaws crested, about its neck a golden sheen, purple on its hinder parts, except the tail on which a brilliant blue is mixed with rose-coloured feathers.

Hortus Sanitatis, bk. iii. § 48.

NEVER man was known to see him feeding. In Arabia he is held a sacred bird, dedicated unto the sun; he liveth 660 years. And the first thing that the young new Phœnix doth is to perform the obsequies of the former Phœnix, late deceased; to translate and carry away his whole nest into the city of the Sun, near Panchaea, and to bestow it full devoutly there upon the altar.

Holland's Pliny, bk. x. ch. ii.

IN the city Chora, there is one of these [date-] trees that bears dates like to pearls, and the bird Phœnix is supposed to have taken name of this date-tree, for it was assured to me that the said bird died with that tree, and revived of itself as the tree sprung again. *Ibid.*, bk. xiii. ch. iv.

Pig (*i.e.*, Sucking-pig).

COMEDY OF ERRORS, i. 2, 44.
TITUS ANDRONICUS, iv. 2, 145.

PIG:—A little young swine.

Minsheu's Dictionary, *s.v.*

THERE were three Sucking-pigs serv'd up in a dish,
Ta'en from the sow as soon as farrowed,
A fortnight fed with dates and muscadine,
That stood my master in twenty marks apiece,
Besides the puddings in their bellies.

Massinger, "The City Madam," ii. 1.

FIVE shillings a Pig is my price at least; if it be a Sow-pig, sixpence more.

Ben Jonson, "Bartholomew Fair," ii. 2.

ROASTED with fire of Juniper and Rosemary-branches.

Ibid., iii. 2.

SOME of my country-men at Wittenburg, desiring to eat a Pig, hardly bought one for half a dollar, and were our selves forced to kill, dress and roast it.

Fynes Moryson, "Itinerary," part iii. p. 84.

V. Boar, Swine.

Pigeon.

As You Like It, i. 2, 99; and iv. 1, 150.
ii. King Henry IV., v. 1, 16.

V. Dove.

To boil Pigeons in black broth.

Dawson, "The Good Huswife's Jewel."

THE sparrow hawk is a fierce enemy to all Pigeons, but they are defended of the kestril, whose sight and voice the spar-hawk doth fear, which the Pigeons or doves know well enough; for where the kestril is, from thence will not the Pigeons go (if the spar-hawk be nigh) through the great trust she hath in the kestril her defender.

Lupton, "A Thousand Notable Things," bk. x. § 3.

PIGEONS do so love the kestril, that if one close young kestrils in a pot, and stop and cover the same close, and shall hang them in four corners of the dove-house, it will procure such a love to the Pigeons of that place, that for the desire of them being so enclosed in the said pots, they will never change that place, so much they love the kestrils their friends after their death.

Lupton, "A Thousand Notable Things," bk. i. § 46.

THAT Pigeons be not hunted or killed of cats at the windows at every passage and at every Pigeon's hole, hang or put little branches of rue; for rue hath a marvellous strength against wild beasts. *Ibid.*, bk. iii. § 38.

IF the skull of an aged man be hanged in a dove-house, Pigeons will be increased there, and will live quietly.

Ibid., bk. viii. § 23.

PIGEONS, now an hurtful fowl by reason of their multitudes, and number of houses daily erected for their increase (which the bauers of the country call in scorn alms-houses, and dens of thieves and such like) whereof there is great plenty in every farmer's yard.

Holinshed, "Description of England," p. 223.

Pike.

ii. KING HENRY IV., iii. 2, 356.

V. Luce.

To boil a Pike with oranges (a banquet dish):—[The Pike was boiled in a pint of water, and a pint of wine, with oranges, dates, spices, sweet butter, and served with its head cut off and placed erect before its body, and an orange in the mouth.

The second part of the "Good Huswife's Jewel," p. 22.]

[Pike was also baked ("Widow's Treasure").]

Pilchard.

TWELFTH NIGHT, iii. 1, 40.

THE inhabitants of Cornwall make great gain by the fishing of Pilchards, which they salt and dry in the smoke, and export an huge multitude of them yearly into Spain and Italy.

Fynes Moryson, "Itinerary," part iii. p. 136.

PILCHARDS otherwise called fumadors, taken on the shore of Cornwall from July to November, saleable in France.

Nashe, "Lenten Stuff."

[*Minsheu* (Dictionary, *s.v.*) seems to confuse Pilchards with sardines.]

[IN Peru] because the maize will not grow, except it first die, they set one or two Pilchards' heads therewith, and thus it groweth abundantly.

Purchas, "Pilgrims," p. 873 (ed. 1616).

Pine.

MERCHANT OF VENICE, iv. 1, 75.

IN the islands of Germany of the Pine-apple tree [*i.e.*, the Pine-tree] cometh dropping and oozing, which is made hard with coldness or with heat, and so turneth into a precious stone that hight *electrum* [*i.e.*, amber]. Also this tree is good to all thing that is kept and continued thereunder, as the fig-tree grieveth and noyeth all things that is thereunder. Also the Pine-tree and alloren-tree [*i.e.*, alder], heled [*i.e.*, covered] with earth deep under the ground, dure and last long time. Pipes and conduits made of Pine-tree, and laid deep under the earth, dure many years. And dureth in an house long time, and rotteth not soon, neither is worm-eaten, but if it be corrupt with dropping of rain.

Bartholomew (*Berthelet*), bk. xvii. § 121.

Pink.

ROMEO AND JULIET, ii. 4, 61.

V. Gilliflower.

Pippin.

ii. KING HENRY IV., v. 3, 2.

V. Apple.

Pismire.

i. King Henry IV., i. 3, 240.

V. Ant.

Plantain.

No salve, sir, but a plantain.
Love's Labour's Lost, iii. 1, 75.

Your plantain-leaf is excellent . . . for your broken shin.
Romeo and Juliet, i, 2, 52.

Weybread hight *Plantago* [Plantain]. It healeth sore wounds and biting of wood hounds, and is contrary to venom, and namely [especially] to the venom of an attercop [spider]. *Bartholomew* (*Berthelet*), bk. xvii. § 129.

V. Spider and Toad.

It is certainly and constantly affirmed that on Midsummer Eve, that is the day before the Nativity of Saint John Baptist, there is found under the root mugwort a coal, which preserves or keeps them safe from the plague, carbuncle, lightning, the quartan ague, and from burning, that bears the same about them. And Mizaldus the writer hereof saith that he doth hear that it is to be found the same day under the root of Plantain; which I know to be of a truth, for I have found them the same day under the root of Plantain, which is especially and chiefly to be found at noon. *Lupton*, "Notable Things," bk. i. § 59.

Plum-tree.

ii. King Henry VI., ii. 1, 97.

Of the Plum-tree is many manner of kind; but the Damascene [Damson] is the best that cometh out of Damask. Only of this tree droppeth and cometh glue and fast gum. Physicians say that it is profitable to medicine, and for to make ink for writers' use.

Bartholomew (*Berthelet*), bk. xvii. § 125.

There are divers sorts of Plums, the Damson, the Apricot, the Pear-Plum, the Wheaten Plum, the Levant Plum, the White Shrag, the Bullace, the Sloes, the Snages, besides other strange Plums that grow in other countries to us unknown.

Batman's addition to *Bartholomew*, bk. xvii. § 125.

[Tusser, at the end of "January's Abstract," mentions besides some of the above Plums, the Cornet-Plum, and Green or Grass-Plums.

References to Plum-broth, Plum-porridge, or Plum-pottage, or, in other words, Plum-pudding, are common enough.]

Pole-cat.

Merry Wives of Windsor, iv. 1, 29.

The Pole-cat stinks very badly, especially when it is angered. Like the badger it has short legs on the left side, and longer ones on the right.

Hortus Sanitatis, bk. ii. § 113.

Pome-water.

Love's Labour's Lost, iv. 2, 4.

V. Apple.

Poor-John.

TEMPEST, ii. 2, 28.

[Nares (Glossary, *s.v.*) says that Poor-John was hake salted and dried, and derives it from *pauvre Jean*, in preference to *pauvre gens* (*Malone*).]

THE French carry into Italy dry fish called Poor-John (brought to them by the English).

Fynes Moryson, "Itinerary," part iii. p. 134.

A SLOP of a rope-hauler is first broken to the sea in the herring-man's skiff or cock-boat, where [he learns] to eat Poor-John out of smutty platters, when he may get it, without butter or mustard. *Nashe*, "Lenten Stuff."

THIS is a patent for the taking of Poor-John and barrel-cod alive and so to preserve 'em in salt-water for the benefit of the fish-mongers. *Brome*, "The Court-beggar," v. 2.

WE thy old friends to thee unwelcome are,
Poor-John and apple-pies are all our fare.
No salmon, sturgeon, oysters, crab, nor conger.

Sir John Harington, "Epigrams," bk. ii. 50.

Poperin-pear.

ROMEO AND JULIET, ii. 1, 38.

V. Pear.

Popinjay.

i. KING HENRY IV., i. 3, 59.

V. Parrot.

POPINJAY or parrot.

Minsheu's Dictionary, *s.v.*

THE parrot, the Popinjay, Philip-sparrow, and the cuckoo.

Nashe, "Lenten Stuff"; and *cf.* *Nares'* Glossary.

Poppy.

OTHELLO, iii. 3, 330.

THIS seed is good to season bread with.

Gerard's "Herbal," *s.v.*

POPPY and mandragora. *Hortus Sanitatis*, bk. i. § 328.

COMMONLY sown it is with coleworts, purslane, rocket and lettuce. *Holland's Pliny*, bk. xix. ch. viii.

Porcupine, Porpentine.

HAMLET, i. 5, 20.

ITS anger is most quick to revenge, so that very often it looses its spines from its back, and wounds dogs or men that are near it. *Hortus Sanitatis*, bk. ii. § 73.

THE Porcupines come out of India and Africa; a kind of urchin or hedge-hog they be; armed with pricks they be both, but the Porcupine hath the longer sharp-pointed quills, and those when he stretcheth his skin, he sendeth and shooteth from him. *Holland's Pliny*, bk. viii. ch. xxxv.

THE pilgrims that come yearly from St. James of Compostella in Spain do bring back generally one of these quills in their caps. The pace of this beast is very slow and troublesome unto it. It is a filthy beast, smelling rank because it liveth so much in the earth; being wild it never drinketh, and I think it eateth apples, roots and rinds of trees, and peradventure snails and such reptile creatures. If men scrape their teeth with their quills, they will never be loose. *Topsell*, "Four-footed Beasts," pp. 457-8.

Potato.

MERRY WIVES OF WINDSOR, v. 5. 21.

[Potatoes were held to be incentives to venery (*vide* Collins' long note at the end of "Troilus and Cressida" in *Malone* and *Steevens*' Shakespeare, vol. xi. ed. 1793).

Potato-pies are mentioned in *Ben Jonson's* "Every Man out of his Humour," *Heywood's* "The English Traveller," and *Dekker's* "Gull's Hornbook," etc. In the "Good Huswife's Jewel" is a recipe for a "potato-tart"; "potatoes marrowed" in *Massinger's* "Guardian," ii. 2).

Potatoes were sometimes cheap, so "Histrio - mastix," ii. 1, 76:

Merchant's Wife: Ha' ye any Potatoes?
Seller: The abundance will not quit-cost the bringing.]

Pricket.

LOVE'S LABOUR'S LOST, iv. 2, 12, etc.

A BUCK is, the first year, a fawn; the second year, a Pricket; the third year, a sorrel; the fourth year, a sore; the fifth, a buck of the first head; the sixth year, a complete buck. "The Return from Parnassus," ii. 5.

Primrose.

WINTER'S TALE, iv. 4, 122

V. Oxlip, Cowslip.

Provençal Roses.

HAMLET, iii. 2, 288.

OF Provence Roses there were various kinds; *e.g.*, the Red Rose, the Damask Rose, and the Great Rose, or Rose of Holland. *Gerard's* "Herbal," *s.v. Rose, q.v.*

Prune.

[As to stewed Prunes, they were usual refreshments in houses of evil repute (i. "King Henry IV.," iii. 3, 128; "Merry Wives of Windsor," i. 1, 296; "Measure for Measure," ii. 1, 92, etc.); but prunes were also used by respectable people ("Winter's Tale," iv. 3, 51).

Damask Prunes (*Lilly*, "Mother Bombie," iii. 4) used in porridge were dried damsons.

Prunes were made into tarts ("The Good Huswife's Treasury," p. 7).]

Pumpion.

This unwholesome humidity, this gross, watery pumpion.
MERRY WIVES OF WINDSOR, iii. 3, 44.

PUMPIONS strangely hate oil, and love water.
Hortus Sanitatis, bk. i. § 352.

THE fruit of Pompions or melons boiled in milk and buttered is a good wholesome meat for man's body. The flesh or pulp of the same sliced and fried in a pan with butter is also a good and wholesome meat; but baked with apples in an oven, it is food utterly unwholesome for such as live idly, but unto robustious and rustic people nothing hurteth that filleth the belly. *Gerard's* "Herbal," *s.v.*

[N.B.—Gerard gives six illustrations of "melons or Pompions," of which some resemble vegetable marrows, and others water-melons.]

Courtier. A GROAT
My ordinary in Pompions baked with onions.
Peregrine. Do such eat Pompions?
Doctor. Yes, and clowns musk-melons.
Brome, "The Antipodes," iv. 5.

FOR ought I see Pompions are as good meat [as asparagus] for such a hoggish thing as thou art.
Ibid., "The Sparagus Garden," iii. 8.

Puppy.

TWO GENTLEMEN OF VERONA, iv. 4, 3.

V. Whelp.

Purple (plant).

HAMLET, iv. 7, 171.

[The Purple orchis (*orchis mas*), which, according to Gerard, Pliny, and other authorities, has certain wonderful properties, derived by the doctrine of signatures from the shape of the root, to which also the "grosser name" is to be ascribed.]

Puttock.

ii. KING HENRY VI., iii. 2, 191.
TROILUS AND CRESSIDA, v. 1, 68.

A BUZZARD, a glede, Puttock or kite.
Minsheu's Dictionary, *s.v. Buzzard.*

[In the passage in ii. "King Henry IV." the Puttock is evidently a kite, as also in *Spenser's* "Faery Queene," v. xii. 30 (quoted in *Nares'* Glossary).]

Quail.

TROILUS AND CRESSIDA, v. 1, 57.
ANTONY AND CLEOPATRA, ii. 3, 37.

[Latin, *Coturnix* (*Minsheu's* and *Cooper's* Dictionaries).]

CURLEWS hight *coturnices*, and hath that name of the sound of the voice. These birds have guides and leaders as cranes have; and, for they dread the goshawk, they be busy to comfort the leaders. Only those birds have the falling-evil as a man hath, and the sparrows also. And they pass the sea, and, when they be weary, they fall down upon the water, and rest upon the one wing, and maketh his sail of the other wing. His best meat is venomous seeds and grains, and for that cause in old time men forbade eating of them; and an herb that hight hellebore is curlews' meat, and if another beast eateth it in great quantity, it is perilous and poison. For beasts have broad and wide veins, by which the smoke passeth, and by strength of that herb, the heart is suddenly cooled and dead; and curlews have strait veins about the heart, and therefore venomous smoke hath no true passage, but he bideth in the stomach, and is there defied [digested] and made subtle, and so it grieveth them not. And he runneth upon the earth most swiftly. And such birds love birds of their own kind.

Bartholomew (*Berthelet*), bk. xii. § 7.

As touching Quails, they always come before the cranes depart. The manner of their flying is in troops; but not without some danger of the sailors, when they approach near to land. For oftentimes they settle in great number on their sails, and there perch, which they do evermore in the night, and with their poise bear down barks and small vessels, and finally sink them. When the south wind blows, they never fly. The foremost of them, as he approacheth near to land, payeth toll for the rest unto the hawk, who presently for his welcome preyeth upon him. Whensoever at any time they are upon their remove and departure out of these parts, they persuade other birds to bear them company. If a contrary wind should arise and drive against them, and hinder their flight—to prevent this inconvenience, they be well provided; for they fly well ballasted either with small, weighty stones within their feet, or else with sand stuffed in their craw.

Holland's Pliny, bk. x. ch. xxiii.

[On the passage in "Antony and Cleopatra," Douce (Illustrations, vol. ii. p. 86-7) gives a note on the classical Quail-fighting. Shakespeare probably got the idea from *North's* "Plutarch."

In "Troilus and Cressida" (*loc. cit.*) "Quail" is a prostitute, because the Quail was supposed to be very salacious (so *Glapthorne's* "Hollander," i. 1).

Quails were boiled ("Good Huswife's Jewel").]

Quince.

ROMEO AND JULIET, iv. 4, 2.

QUINCES are seldom eaten raw; being roasted or baked they are more pleasant. The woman with child, which eateth many Quinces during the time of her breeding, shall bring forth wise children and of good understanding. The marmalade or cotiniate [is] made of Quinces and sugar. Many other excellent, dainty and wholesome confections are to be made of Quinces, as jelly of Quinces, and such odd conceits. *Gerard's* "Herbal," *s.v.*

MANY use syrup of Quinces at the second course after wine, and it prevents drunkenness.

Hortus Sanitatis, bk. i. § 118.

[Recipes for preserving Quinces are given in the second part of the "Good Huswife's Jewel," and in the "Widow's Treasure." Quince-cakes and marmalade are mentioned in *Massinger's* "New Way to Pay Old Debts," ii. 2.]

Rabbit.

LOVE'S LABOUR'S LOST, iii. 1, 19.

V. Coney.

Rabbit-sucker.

i. KING HENRY IV., ii. 4, 480.

[A young or sucking Rabbit.]

Radish.

i. KING HENRY IV., ii. 4, 206.
ii. KING HENRY IV., iii. 2, 334.

OF the seed of the rape, and also of seed of Radish is oil made, that is needful in many uses, and namely in lamps. *Bartholomew* (*Berthelet*), bk. xvii. § 137.

THE Radish is hostile to vines; for though they be sown round them, they [the vines] turn away from them, because of repugnance of nature. It helps against the bitings of vipers. With honey it restores hair in baldness.

Hortus Sanitatis, bk. i. § 384.

WE will have a bunch of Radish, and salt, to taste our wine.

Ben Jonson's "Every Man in his Humour," i. 5, *ad fin.*
So *Randolph*, "The Jealous Lovers," iii. 5.

Raisin.

Raisins o' th' sun.
WINTER'S TALE, iv. 3, 26.

RAISIN is made in many manner wise. For sometime the stalk thereof is woven and wound, and so the grape in certain days is for-dried by heat of the sun. And this is best to eat. And sometime the grapes be wounden in vine-leaves, and be bound with thread, for the grapes should not seed, and be put in an oven so bound and wrapped, and be dried when the heat is temperate. In such manner sometime Raisins be made in chimneys.

Bartholomew (*Berthelet*), bk. xvii. § 183.

[The first mode of preparation will produce "raisins of the sun."]

ALMONDS and Raisins.

Anthony Munday, "English Roman Life," ch. iii. (1590).

MALAGA Raisins to make him long-winded.

Webster's "Devil's Law Case," iv. 1.

Ram.

AS YOU LIKE IT, iii. 2, 87.

THE Ram is a beast that beareth wool, pleasing in heart, and mild by kind, and is duke, prince and leader of sheep. Therefore kind giveth him great strength passing other sheep. The Ram hath a worm in his head, and for fretting of that worm and itching, the Ram is excited and busheth full strongly, and smiteth full hard all that he meeteth. He is more cruel in heart than the ewes, and his cruelness abateth if his horns be pierced nigh to the ear. If his right gendering stone be bound, he gendereth females, and if the left be bound, he gendereth males, and he gendereth males in the Northern wind, and females in the Southern wind. And such Rams as have black veins under their tongues, such lambs they gender in colour, and if they be white, the lambs are white, and if they be speckled, the lambs be so. The Ram hath a full hard forehead, nigh as horn, and feeble temples, and somewhat gristly. And when it rains, they flee not the rain until they be dead. And they dread kindly the thunder, as sheep do. And they sleep with the sheep tofore midnight, and after part, and change, and turn from side to side in sleeping; for from springing-time to harvest, they sleep on the one side, and then unto springing-time they sleep on the other side; and hold up their heads while they sleep, except they be sick; and they chew their cud sleeping, as they do waking. And if it happen that they stray and go away, they come not again, but if the herd bringeth them again. And their flesh is better than flesh of lambs and of ewes.

Bartholomew (*Berthelet*), bk. xviii. § 3.

A COMB made of the right horn of a Ram doth take away the head-ache being on the right side of the head, if the pained head be combed therewith. If the pain be on the left side of the head, then a comb made of the left horn of a Ram doth take it away. This I had out of an old book. *Lupton*, "Notable Things," bk. iv. § 4.

THEY were wont to hang a shrimp at the horn of the Ram, and then the wolf will never set upon their flocks. If the horns of a Ram be buried in the earth, they will turn into the herb spirage [? misprint for spinach, or for spurge]; for rottenness and putrefaction is the mother of many creatures and herbs.

Topsell, "Four-footed Beasts," p. 493.

V. Lamb, Sheep.

Rat.

The very rats
Instinctively have quit it.
TEMPEST, ii. 1, 147.

THEIR tail is counted venomous. The younger Rats bring food unto the elder, because they are grown to a great and unwieldy stature of body. With the dung of

Rats, the Physicians cure the falling-off of the hair. And if their urine do fall upon the bare place of a man, it maketh the flesh rot unto the bones, neither will it suffer any scar to be made upon the bones.

Topsell, "Four-footed Beasts," pp. 403-4.

It is found by observation that Rats and Dormice will forsake old and ruinous houses, three months before they fall; for they perceive by an instinct of nature, that the joints and fastening together of the posts and timber of the houses, by little and little will be loosed, and so thereby that all will fall to the ground.

Lupton, "Notable Things," bk. ii. § 87.

It is said that no Rats have ever been seen in this town [Hatfield, Yorkshire]. *Camden,* "Britannia," col. 849.

V. **Mouse, Water-rat, Island.**

Ratsbane.

ii. King Henry IV., i. 2, 48.

[The following quotation from *Holland's Pliny,* bk. xxii. ch. xviii., seems appropriate:

"If there be water and oil mingled to the juice of Chameleon [carline Thistle], it draweth rats and mice to it, but it is their bane, unless presently they drink water."

But probably Shakespeare used the word as simply meaning any poison. *Florio* gives "Ratsbane" as an equivalent for corrosive sublimate and for arsensic.]

Raven.

The Raven beholdeth the mouth of her birds when they yawn. But she giveth them no meat ere she know and see the likeness of her own blackness, and of her own colour and feathers; and when they begin to wax black, then afterward she feedeth them with all her might and strength. Ravens' birds be fed with dew of heaven all the time that they have no black feathers. And is a crying fowl, and hath diverse sound and voice; for among fowls only the Raven hath four and sixty changings of voice. And is a guileful bird, and taketh away things thievishly, and layeth and hideth them in privy places.

And they have many birds, and, for they be many, they throw away some of their birds,—for fowls which have many birds throw away some of them. Also the black Raven fighteth with the ass, and with the bull, and flieth upon them, and grieveth them, and smiteth with the bill, and smiteth out their eyes. Also the black Raven is friend to the fox, and therefore he fighteth with the brock and with other small beasts to help the fox.

Bartholomew (*Berthelet*), bk. xii. § 10.

THEY are said to conceive and to lay their eggs at the bill. The young become black on the seventh day. The Raven is stronger by day, and the owl by night, and they eat one another's eggs by turns. It abstains from drinking so long as the fig-tree rejoices in the sweetness of its fruit.

Hortus Sanitatis, bk. iii. § 34.

IF women great with child chance to eat a Raven's egg, they shall be delivered of their children at the mouth.

Holland's Pliny, bk. x. ch. xii.

IF a Raven's eggs be boiled and put again in the nest, straightway the Raven goes to a certain island in the Red Sea, where Aldoricus or Alodrius is buried, and brings a stone with which it touches its eggs, and immediately they become raw as they were before. It is wonderful that boiled eggs should be revived. Now if that stone be set in a ring with a laurel-leaf under it, and a man bound with chains, or a closed gate be touched [therewith], straightway the bound shall be loosed, and the gate be opened. And if that stone be put in the ear, it gives understanding of all birds. This stone is of diverse colours, and causes all anger to be forgotten.

Albertus Magnus, "Of the Virtues of Animals."

Red-breast, Robin Redbreast.

TWO GENTLEMEN OF VERONA, ii. 1, 22.

THE bird which is named Robin or Redbreast in winter, the same is Red-tail [or Red-start] all summer long.

Holland's Pliny, bk. x. ch. xxix. So *Minsheu's* Dictionary, *s.v. Robin-redbreast* and *Red-start*.

A ROBIN REDBREAST finding the dead body of a man or woman will cover the face of the same with moss; and as some hold opinion, he will cover also the whole body.

Lupton, "Notable Things," bk. i. § 37.

V. Ruddock.

Reremouse, Rearmouse.

MIDSUMMER NIGHT'S DREAM, ii. 2, 4.

V. Bat.

Rhinoceros.

MACBETH, iii. 4, 101.

RHINOCEROS, the unicorn, is a wild beast by kind, and may not be tamed in no wise; and if it hap that he be taken in any wise, he may not be kept in no manner; for he is so unpatient and so angry that he dieth anon.

Bartholomew (*Berthelet*), bk. xviii. § 90.

A RHINOCEROS,—his hide or skin of the colour of the box-tree; an enemy to all beasts of rapine and prey, as the lion, leopard, bear, wolf, tiger, and the like; but to others as the horse, ass, ox, sheep, etc.—which feed not upon the life and blood of the weaker, but of the grass and herbage of the field,—harmless and gentle, ready to succour them, when they be any way distressed.

Thos. Heywood, "London's Gate to Piety" (1638).

ALL the later Physicians do attribute the virtue of the Unicorn's horn to the Rhinoceros' horn, but they are deceived. *Topsell*, "Four-footed Beasts," p. 463.

V. Unicorn.

Rice.

WINTER'S TALE, iv. 3, 11.

THE plants of Rice did grow in my garden. In England we use to make with milk and Rice a certain food or pottage. Many other good kinds of food is made with this grain, as those that are skilful in cookery can tell.

Gerard's "Herbal," *s.v.*

Roe (or Roebuck).

TAMING OF THE SHREW, *Induction*, 2, 50.

THEIR swiftness doth not only appear upon the earth, but also upon the waters, for with their feet they cut the waters when they swim as with oars. It hath also been believed that a Roe doth not change her horns, because they are never found; whereas in truth, they fall off yearly as doth a Hart's, but they hide them to the intent they should not be found. They never wink, no, not when they sleep. They are often taken by the counterfeiting of their voice, which the hunter doth by taking a leaf and hissing upon it.

Topsell, "Four-footed Beasts," pp. 91-2.

V. Hart.

Rook.

MACBETH, iii. 4, 125.

A ROOK, Chough or Daw.

Minsheu's Dictionary, *s.v.*

THE crow liveth not altogether of carrion, for the Rook eateth of other food. The Crows and Rooks have a cast by themselves, for when they meet with an hard nut which they be not able to crack, they will fly aloft and fling it against some rock or tile-house once or twice, yea, and many times together, till it be so crushed and bruised, that they may easily break it quite.

Holland's Pliny, bk. x. ch. xii.

V. Crow.

Rope.

TEMPEST, i. 1, 33.

IF you take the Rope with which a thief is or has been hanged, and some of the straw which is carried into the air by an eddy of wind, and put it in a pot, and put that pot with others,—that pot will break all the others. Also take a piece of the aforesaid Rope and put it in the instrument with which bread is put into the oven, and when he whose duty it is to put it into the oven wants to put it in he will not be able to do so, but it will fly out.

Albertus Magnus, "Of the Wonders of the World."

Rose.

AMONG all flowers of the world, the flower of the Rose is chief, and beareth the price. And therefore oft the chief part of man, the head, is crowned with flowers of Roses. Of green Roses *aqua rosacea* [rose-water] is made by seething of fire, or of the sun, and this water is good in ointment for ladies, for it cleanseth away webs and foul specks of the face, and maketh the skin thin and subtle. Powder of dry Roses comforteth wagging teeth that be in point to fall.

Bartholomew (*Berthelet*), bk. xvii. § 136.

DODONEUS writeth of ten kind of Roses, among the which the Eglantine Rose, and Musk-Rose, yellow and white. There is one Rose growing in England is worth all these *Rosa sine spina* [by which he seems to mean Queen Elizabeth, and he breaks off into a discourse to the other flowers on self-indulgence, pillage of the clergy, oppression of the poor, etc.].

Batman's addition to *Bartholomew*, *loc. cit.*

[Gerard ("Herbal," *s.v.*) describes the following sorts of Roses: the White, the Red, the Provence or Damask, the Rose without prickles, the Holland or Provence, the Single and Double Musk-Rose, the Great Musk-Rose, the Velvet, the Yellow, the Double Yellow, the Double and Single Cinnamon. Of wild Roses: the Eglantine or Sweetbriar, the Double Eglantine, the Briar or Hip-tree, and the Pimpernell or Burnet. He saith further that "the distilled water of Roses being put into junketting dishes, cakes, sauces and many other pleasant things giveth a fine and delectable taste. The making of the crude or raw conserve is very well known, as also sugar roset, and divers other pretty things made of sugar and Roses, which are impertinent unto our history, because I intend neither to make thereof an apothecary's shop, nor a sugar-baker's storehouse, leaving the rest for our cunning confectioners."

Rose-water was used to wash in ("Taming of the Shrew," Induction, 1, 57), and to mix with wine. "A cheater meeting a stranger in the dark gets him by some sleight to a tavern, where calling for two pints of sundry wines, the drawer setting the wines down with two cups, as the custom is, the Jumper tastes of one pint (no matter which) and finds fault with the wine, saying 'tis too hard, but Rose-water and sugar would send it down merrily; and for that purpose takes up one of the cups, telling the stranger he is well acquainted with the boy at the bar, and can have twopennyworth of Rose-water for a penny of him, and so steps from his seat, the stranger suspecting no harm, because the fawn-guest leaves his cloak at the end of the table behind him. But this Jump [swindle] coming to be measured, it is found that he that went to take his rising at the bar hath stolen ground and out-leaped the other more feet than he can recover in haste, for the cup is leaped away with him, for which the woodcock that is taken in the springe must pay fifty shillings or three pound, and hath nothing but an old threadbare cloak not worth 10 groats, to make amends for his losses" (*Dekker's* "Bellman of London").

Rose-water was also put in mince-pies ("Good Huswife's Treasury," p. 4).

TAKE the seed of a Rose, and the seed of mustard, and the foot of a weasel, and hang these on a tree, and from thenceforth it will bear no fruit. And if the aforesaid be put upon a net, the fish will collect there. And if the said dust be put in a lamp, and then it be lighted, all men will seem to be as black as the devil. And if the said powder be mixed with olive-oil and quick sulphur, and a house be smeared with this while the sun is shining, it will appear to be all on fire.

Albertus Magnus, "Of Virtues of Herbs," § 15.

Rosemary.

There's rosemary, that's for remembrance.
HAMLET, iv. 5, 175.
So WINTER'S TALE, iv. 4, 74.

THEY make hedges of it in the gardens of England, being a great ornament unto the same. Rosemary is spice in the German kitchens. The flowers, made up into plates with sugar after the manner of sugar-roset and eaten, comfort the heart and make it merry, quicken the spirits and make them more lively. *Gerard's* "Herbal," *s.v.*

[Rosemary was used both at weddings and funerals. "There will be charges saved too; the same Rosemary that serves for the funeral will serve for the wedding" (*Middleton*, "The Old Law," iv. 1, 36). So also *Herrick's* "Hesperides," "The Rosemary Branch":

> Grow for two ends, it matters not at all
> Be 't for my bridal, or my burial.

From the same author we learn that the Rosemary was gilded:

> My wooing's ended; now my wedding's near,
> When gloves are given, gilded be you there.
>
> ("To Rosemary and Bays.")]

THE last of the flowers is the Rosemary (*Rosmarinus*, the Rosemary is for married men), the which by name, nature. and continued use, man challengeth as properly belonging to himself. It over-toppeth all the flowers in the garden, boasting man's rule. It helpeth the brain, strengtheneth the memory, and is very medicinable for the head. Another property of the Rosemary is, it affects the heart. Let this *Rosmarinus*, this flower of men, ensign of your wisdom, love and loyalty, be carried not only in your hands, but in your heads and hearts.

Roger Hackett, "A Wedding Present," quoted in *Brand's* "Popular Antiquities," vol. ii. 49.

THE Rosemary that was washed in sweet water to set out the bridal is now wet in tears to furnish her burial.

Thos. Dekker, "The Wonderful Year 1603."

THE price of flowers, herbs and garlands rose wonderfully [during the plague-time], insomuch that Rosemary which had wont to be sold for 12 pence an armful went now for six shillings a handful.

Ibid.

AND stuck her with Rosemary to sweeten her; she was tainted ere she came to my hands.

Middleton, "The Old Law," iv. 1, 12.

[Before a wedding.] Let us dip our Rosemaries
In one rich bowl of sack to this brave girl
And to the gentleman.

Jasper Mayne, "The City Match," v. 1 (1639).

Ruby.

Measure for Measure, ii. 1, 101.

Among these red gems, the Rubies otherwise called carbuncles challenge the principal place.

Holland's Pliny, bk. xxxvii. ch. 7.

[So *Minsheu's* Dictionary, "Ruby, *v.* Carbuncle," therefore we may suppose that the stones were considered to be identical.]

V. Carbuncle.

Ruddock.

With fairest flowers,
While summer lasts, and I live here, Fidele,
I'll sweeten thy sad grave; thou shalt not lack
The flower that's like thy face, pale primrose, nor
The azured harebell, like thy veins, no, nor
The leaf of eglantine, whom not to slander,
Out-sweeten'd not thy breath: the ruddock would
With charitable bill,—O bill, sore-shaming
Those rich-left heirs that let their fathers lie
Without a monument!—bring thee all this;
Yea, and furr'd moss besides, when flowers are none,
To winter-ground thy corse.

Cymbeline, iv. 2, 218-29.

Call for the robin-redbreast and the wren
Since o'er shady groves they hover,
And with leaves and flowers do cover
The friendless bodies of unburied men;
Call unto his funeral dole
The ant, the field-mouse, and the mole
To rear him hillocks that shall keep him warm
And (when gay tombs are robb'd) sustain no harm;
But keep the wolf from thence, that's foe to men,
For with his nails he'll dig them up again.

Webster, "White Devil, or Vittoria Corombona," Act. v.

V. Redbreast.

Rue.

HAMLET, iv. 5, 181-2.

RUE is a medicinable herb, and is full fervent. Weasels teach that this herb is contrary to venom and to venomous beasts, for he eateth first Rue, and balmeth himself with the smell and the virtue thereof, ere he fighteth with the serpent, and fighteth afterward sicherly [safely], and reseth [rusheth] on the cockatrice, and slayeth him. Rue eaten raw sharpeth the sight of the eyes; and Rue, eaten or drunk withstandeth mightily all venom and biting of venomous beasts, if it be stamped with salt, garlic and nuts, and healeth wonderly such bitings, and the smell of Rue driveth and chaseth away all venomous beasts out of gardens, and is therefore planted about sage to drive away serpents and toads, which love sage best. And Rue hateth winter, dung and humour, and thriveth well in dry weather. Ashes should be meddled with seeds thereof, to destroy malshrags [caterpillars] and other worms.

Bartholomew (*Berthelet*), bk. xvii. § 141.

RUE, or Herb-grace.

Gerard's "Herbal," *s.v.*; so also *Minsheu's* Dictionary.

IT is a common received opinion that Rue will grow the better if it be filched out of another man's garden; and it is as ordinary a saying, that stolen bees will thrive worst.

Hollands Pliny, bk. xix. ch. vii.

Is a man disposed to drink freely, and to sit squarely at it? Let him before he begin take a draught of the decoction of Rue-leaves, he shall bear his drink well, and withstand the fumes that might trouble and intoxicate his brains. *Ibid.*, bk. xx. ch. xiii.

WHAT savour is better, if physic be true,
For places infected, than wormwood and Rue?
It is as a comfort, for heart and the brain,
And therefore to have it, it is not in vain.

Tusser, "Five Hundred Points," July's "Husbandry," st. 11.

TAKE a little Rue or herb-grace, and stamp the same, then strain out the juice thereof, and after you have thus done, let the party, that is pained with the toothache, drop three or four drops of the same juice into his ear, on that side the pain is, and let him lie on his other side an hour or two, and it will not only take away the present pain, but also the party that trieth it shall never be troubled with the tooth-ache afterward. This was reported unto me for very truth by one which had proved the same.

Lupton, "Notable Things," bk. x. § 61 (*bis.*).

COLEWORTS [cabbage] and Rue (otherwise called herb grace) are so contrary in nature the one to the other, that they ought not to be sown nigh together.

Ibid., bk. vi. § 55.

STAMP Rue with oil of roses, and lay the same something thick upon the crown of the head of one that is sick, the same being first shaven, and if the same party do [s]neeze within six hours after, he will escape that sickness. If not, he will die thereof.

Ibid., bk. vii. § 28.

Rush.

TAMING OF THE SHREW, iv. 1, 45.
ROMEO AND JULIET, i. 4.

DRY Rush to kindle fire and lanterns; and this herb is put to burn in prickets [wax candles, *i.e.*, as wick] and in tapers. The rind is stripped off unto the pith, and is so dried, and a little is left of the rind on the one side to sustain the tender pith, and the less is left of the rind, the more clear the pith burneth in a lamp, and is the sooner kindled. Of Rushes be rushen vessels made. And about Memphis and in Ind be such great Rushes, that they make boats thereof. And of Rushes be charters made, in the which were epistles writ, and sent by messengers. Also of Rushes be made paniers, boxes and cases, and baskets to keep in letters and other things. With Rushes water is drawn out of wine. And also they make thereof paper to write with.

Bartholomew (*Berthelet*), bk. xvii. § 126

FIVE kinds of Rushes are written of: the candle Rush, the hard Rush and Fen-rush, the Bull-rush or Mat-rush, squinauth [or, camel's hay].

Batman's addition to *Bartholomew*, *loc. cit.*

[The references to the Rushes strewed on the floors of rooms are so numerous that there is no need to quote them. The rushes used were often sweet-scented ones, which are still found in some marshes in the Eastern Counties. Bulrushes were also used (*Dekker's* "Bellman of London"), and hay (*Hentzner's* "Itinerary"), or flowers: "Strew all my bowers with flags and water-mints" (*Lilly's* "Woman in the Moon," iii. 2). The Rushes must have been frequently changed, for "all the ladies and gallants lie languishing upon the Rushes" (*Ben Jonson*, "Cynthia's Revels," ii. 5), and they helped out conversation: "If you had but so far gathered up your spirits to you, as to have taken up a rush, when you were out, and wagged it, thus, or cleansed your teeth with it" (*ibid.*, iii. 1). As to the price, the cost was 3d. per burthen in 1559 (*Lyte's* "Eton College," p. 169).]

A RUSH dried and put into wine, if there be any water therein, draws it to it (the wine left alone, or together)—which is good and profitable for trying of wine.—*Mizaldus.*

Lupton, "Notable Things," bk. iii. § 77.

Rye.

AS YOU LIKE IT, v. 3, 23.

FOR your brown bread, or bread for your hind-servants, which is the coarsest bread for man's use, you shall take of barley, two bushels, of pease, two pecks, of wheat or Rye, a peck, a peck of malt, etc.

Gervase Markham, "English Housewife's Skill in Baking," p. 187 (1656).

MEALS for bread are either simple or compound; simple, as wheat, and Rye,—or compound, as Rye and wheat mixed together, or Rye, wheat and barley mixed together.

Ibid., p. 185.

Sables.

HAMLET, iv. 7, 81.

THE fur-marten is most excellent, for princes and great nobles are clothed therewith, every skin being worth a French crown, or four shillings at the least.

Topsell, "Four-Footed Beasts," p. 386.

[A thousand ducats were sometimes given for a suit of sables (*Bishop's* "Blossoms," 1577), and none under the degree of an earl might use Sables ("Statute of Apparel," 24 Henry VIII, c. 13, this quotation and the last being from Malone's note on this passage).]

Sack.

[Condensed from *Nares'* Glossary: "Sack, a Spanish wine of the dry or rough kind; *vin Sec*, French; *Sac*" (*Sekt*, which also means a dry champagne), "German. It was spelt 'seck,' and came from Xeres, and therefore was the same as sherry. Gervase Markham mentions other kinds of Sack, as Canary and Malaga" [which he says are stronger, those of Galicia and Portugal being smaller ("English Housewife's Skill in Wines," p. 118, 1656)]. "Sack was the general name for white wines; where sherry was meant, it was distinguished as *sherris Sack* ('Bartholomew Fair,' v. 4). In 'Pasquil's Palinodia' (1619) *Sack* and *sherry* are used throughout as perfectly synonymous:

'Give me Sack, old Sack, boys,
  To make the Muses merry,
The life of mirth, and the joy of the earth,
  Is a cup of good old sherry.'"

But Falstaff generally drank his Sack "brewed," or mulled, or "burnt" ("Merry Wives of Windsor," iii. 5, 1. 30, and ii. 1, 223), or with sugar (i. "King Henry IV.," i. 2, 125, and ii. 4, 515), and perhaps a toast in it ("Merry Wives of Windsor," iii. 5, 3), or eggs (*ibid.*, iii. 5, 31); and the eggs were sometimes rotten (*Heywood's* "Fair Maid of the West," iii. 4, *ad fin*). Ginger was put into mulled Sack (*Beaumont and Fletcher*. "The Captain," iv. 2). Lime was used to adulterate Sack (i. "King Henry IV.," ii. 4, 130, and "Merry Wives of Windsor," i. 3, 10; and *Sir Richard Hawkins'* "Voyages," as quoted by Warburton); and horseflesh was hung in the cask to keep it quick (" *Webster*, "Westward Ho!" iv. 1, and *Glapthorne's* "Hollander").]

My wag answered me, when I struck him for drinking Sack :—"Master, it is the sovereignest drink in the world, and the safest for all times and weathers; if it thunder, though all the ale and beer in the town turn, it will be constant; if it lighten, and that any fire come to it, it is the aptest wine to burn, and the most wholesomest when it is burnt. So much for summer. If it freeze, why it is so hot in operation, that no ice can congeal it; if it rain, why, then, he that cannot abide the heat of it may put in water. So much for winter."

Lilly, "Mother Bombie," ii. 5.

Saffron.

All's Well that Ends Well, iv. 5, 2.
Winter's Tale, iv. 3, 17.

Saffron is sometimes counterfeited with a thing that is called *crocomagina*. *Crocomagina* is called the superfluity of spicery of the which Saffron ointment is made. He that drinketh Saffron first shall not be drunken; and garlands thereof letteth drunkenness, and letteth a man that he may not be drunk. And cureth biting of serpents and of attercops, and stinging of scorpions.

Bartholomew (*Berthelet*), bk. xvii. § 41.

Saffron colour dyeth and coloureth humours and liquors more than citrine, and tokeneth passing heat and distemperance of blood in the liver. Most hottest birds of prey have their utter parts yellow of colour as their feet and bills.

Ibid., bk. xix. § 16.

[Saffron was grown in Essex and Cambridgeshire (*Fynes Moryson's* "Itinerary," part iii. p. 140), and was used to colour not only pies ("Winter's Tale, *ut supra*), but custards (*Heywood's* "Fair Maid of the Exchange") and porridge (*Beaumont and Fletcher*, "Women Pleased," iii. 2), as well as for a dye ("All's Well that End's Well," *loc. cit.*, and Steevens' notes).]

My quaint knave
He tickles you to death, makes you die laughing,
As if you had swallow'd down a pound of Saffron.

Webster, "Vittoria Corombona."

[Sage.]

[Sage is not mentioned by Shakespeare, but the following statements are worth recording]:

THE learned and wise among the Persians affirm, that if Sage be putrefied or laid to rot in horse-dung, while the sun and moon do both occupy the second face of Leo; thereon will breed a bird like an ousel, or black-bird,—the ashes whereof being burned, and strewed or cast into a burning lamp, will make the house seem to be full of serpents. *Lupton*, "Notable Things," bk. iii. § 72.

[SAGE treated as above] generates a certain worm or bird with a tail like a black-bird, and if any one's breast be touched with its blood, he will lose his senses for fifteen [hours? or days?] and more. And if the aforesaid serpent be burnt, and its ashes be thrown in the fire, there will straightway be a horrible clap of thunder. And this has been tried by the moderns.

Albertus Magnus, "Of the Virtues of Herbs."

SAGE is singular good for the head and brain; it quickeneth the senses and memory. *Gerard's* "Herbal," *s.v.*

V. **Toad.**

Salamander.

i. KING HENRY IV., iii. 3, 53.

THE Salamander is like to the newt in shape, and is never seen but in great rain, and faileth in fair weather, and his song is crying. And he quencheth the fire that he toucheth as ice does, and water frore [frozen]. And out of his mouth cometh white matter, and if that matter touch a man's body, the hair shall fall, and what it toucheth is corrupt, and infected, and turneth into foul colour. And is a pestilent beast most venomous, for the Salamander infecteth fruit of trees, and corrupteth water, so that he that eateth or drinketh thereof is slain anon. And if his spittle touch the foot, it infecteth and corrupteth all the man's body. Of all beasts only the Salamander liveth in fire. And a certain kind of Salamander hath rough skin and hairy, as the skin of the sea seal; of the which skin be sometime girdles made for the use of kings; the which girdles when they be full old be thrown into the fire harmless, and without wem [blemish] purged, and as it were renewed, and of that skin be tongues and bonds [wicks] made in lamps and in lanterns that be never corrupt with burning of fire.

Bartholomew (*Berthelet*), bk. xviii. § 92.

If he creepeth on a tree, he infecteth all the apples, and slayeth them that eat thereof, and if he falleth into a pit, he slayeth all that drink of the water.

Ibid., bk. xviii. § 9.

The Salamander naturally loveth milk, and therefore, sometimes in the woods or near hedges, it sucketh a cow that is laid, but afterwards that cow's udder or stock drieth up, and never more yieldeth any milk. It is not bred of the fire as crickets are.

Topsell, "History of Serpents," pp. 747-8.

Salmon.

KING HENRY V., iv. 7, 32.

[It comes well to pass that he begins this course in a year when there is so great plenty of excellent venisons and such store of Salmons, that the like hath not been seen in the Thames these forty years ("Chamberlain's Letters to Sir Dudley Carleton," from *Nichol's* "Progress of King James I.," vol. iii. p. 394).

Half a Salmon cost 9s. and a side of fresh Salmon 8s. in 1573, but,—"You see this Salmon? it cost but sixpence" (*Rowley*, "A Woman Never Vexed," Act. i.).

Salmon was boiled in water with rosemary and thyme, and a quart of strong ale, and a good deal of vinegar was put in the broth ("The Good Huswife's Jewel," part ii. p. 25). Calvered Salmon was a luxury ("The Alchemist," ii. 2).]

Samphire.

KING LEAR, iv. 6, 16.

ROCK-SAMPHIRE groweth on the rocky cliffs at Dover, Winchelsea (by Rye), about Southampton, [and] the Isle of Wight. The leaves kept in pickle, and eaten in salads with oil and vinegar, is a pleasant sauce for meat.

Gerard's "Herbal," *s.v.*

Sapphire.

MERRY WIVES OF WINDSOR, v. 5, 75.

SAPPHIRE is a precious stone, and is blue in colour, and most like to heaven in fair weather and clear, and is best among precious stones, and most precious, and most apt and able to fingers of kings, for it lighteneth the body, and keepeth and saveth limbs whole and sound. In the same veins of Sapphire in the middle is a certain kind of carbuncle found; therefore many men ween that the Sapphire is the carbuncle's mother. And the Sapphire hath virtue to rule and accord them that be in strife, and helpeth much to make peace and accord. Also it hath virtue to abate unkind [unnatural] heat, for the Sapphire cooleth much the heat of burning fevers, if it be hanged nigh the pulse and the veins of the heart. Also it hath virtue to comfort and to glad the heart. His virtue is contrary to venom, and quencheth it every deal. And if thou put an attercop [spider] in a box, and hold a very Sapphire of

Ind at the mouth of the box any while, by virtue thereof the attercop is overcome and dieth, as it were suddenly. And this same I have seen proved oft in many and diverse places. And they that use nigromancy mean that they have answer of god more thereby than by other precious stones. Also witches love well this stone, for they ween that they may work certain wonders by virtue of this stone. This stone bringeth men out of prison bonds, and undoeth gates and bonds that it toucheth. The Sapphire loveth chastity, and therefore lest the effect thereof be let in any wise by his uncleanness that him beareth, it needeth him that beareth it to live chaste. Also this stone doth away envy, and putteth off dread and fear, and maketh a man bold and hardy, and master and victor, and maketh the heart steadfast in goodness, and maketh meek and mild and goodly.

Bartholomew (*Berthelet*), bk. xvi. § 87.

Savory.

WINTER'S TALE, iv. 4, 104.

SUMMER SAVORY maketh thin, and is boiled and eaten with beans, pease, and other windy pulses.

Gerard's "Herbal," *s.v.*

Scorpion.

MACBETH, iii. 2, 36.

A SCORPION is a land worm with a crooked sting in the tail, and it stingeth with the tail, and sheddeth venom in the crooked wound. And it is his property that he smiteth never, nor hurteth never the palm of the hand. And they bring forth small worms shapen as eggs, and breedeth fervent and right pestilential venom as serpents do. And the venom of Scorpions noyeth and grieveth three days full sore, and afterward slayeth with soft death, but it be holpen and succoured the sooner. And the Scorpion smiteth maidens with death's stroke, when he smiteth and stingeth them, and women also, but he smiteth not men so soon; and grieveth most and noyeth in the morrow-tide, when they come out of their dens. The Scorpion's tail is alway ready to smite and sting; and he stingeth and smiteth aslant, and sheddeth in the smiting white venom. Some have [two] stings, and among these Scorpions the males be most grievous, and namely in time of love. And they have certain knots or rivels [wrinkles] in the tail, and the more such they have, the venom is the worse, and they have sometime such knots six or seven. In Africa some Scorpions have feathers [wings], and those be full grievous. And because of winning [*i.e.*, of gain] enchanters gather venom of divers lands, and labour for to bear these winged Scorpions into Italy, but they may not live under heaven within the country of Italy. To a man smitten of the Scorpion, ashes of Scorpions burnt, drunk in

wine, is remedy. Also Scorpions drowned in oil helpeth and succoureth beasts that be stung with Scorpions. The Scorpion hurteth no beast that hath no blood. And some Scorpions breed and bring forth eleven young Scorpions, and the mother eateth them sometime, but one of them that is most sly leapeth on the thigh of the mother, and sitteth there safe and secure from the stinging of the tail, and from the biting of the mouth, and this slayeth his father, and wreaketh the death of his brethren ; and kind ordaineth this provision, for such a pestilential kind should not multiply too much. And some Scorpions do eat some venomous things, and have the worse venom, and so dragons do eat Scorpions, and those be worst.

Bartholomew (*Berthelet*), bk. xviii. § 98.

An Italian, through the oft smelling of an herb called basil, had a Scorpion bred in his brain, which did not only a long time grieve him, but also at the last killed him. Jacobus Hollerius a learned Physician affirms it for truth. Take heed, therefore, ye smellers of basil.

Lupton, "Notable Things," bk. i. § 38.

If any be bitten or stricken of a Scorpion, which shall eat basil the same day, he shall be made whole thereof ["which" refers to the man, not to the Scorpion, which might refuse to eat basil]. *Ibid.*, bk. v. § 66.

One handful of basil with ten sea-crabs, stamped or beaten together, doth make all the Scorpions to come to that place that are nigh to the same. *Ibid.*, § 73.

Alexandrinus Jovianus Pontanus doth say, that he saw a man was grievously stung or stricken of a Scorpion, which presently was delivered and helped thereof, with drinking of frankincense, wherein was sealed the sign of a Scorpion: being after made in powder. But it must be graven in the stone of a ring (Scorpio ascending), the moon then being there, and placed in the Angle, and the frankincense must be sealed with that seal, when the moon is in Scorpio, and found in an Angle. *Ibid.*, bk. ix. § 40.

There is an ancient town in Afric called Pescara, wherein the abundance of Scorpions do so much harm, that they drive away the inhabitants all the summer-time every year until November following. The authors have observed seven several kinds. The fifth kind eateth herbs, and the bodies of men, and yet remaineth insatiable; it hath a bunch on the back, and a tail longer than other Scorpions. The sixth is like a crab, and is of a great body, and hath tongs and takers very solid and strong, like the gramuel or crayfish, and is therefore thought to take the beginning from that fish. The seventh hath wings on the back, like the wings of a locust. They are all little living creatures, not much differing in proportion from the great scarabee or horse-fly, except in the fashion of their tails. The countenance is fawning, and virgin-like; notwithstanding the fair face, it beareth a sharp sting in the tail. And of all other things they love fresh and clean linen, and next to their flesh put on this clean linen, as a man would put on a shirt. The manner of their breed or generation is double,—one way is by putrefaction, and the other by laying of eggs. When sea-crabs die, and their bodies are dried upon the earth, when the sun entereth into Cancer and Scorpio, out of the putrefaction thereof ariseth a Scorpion, and so out of the putrefied body of the crayfish burned, and out of the basilisk beaten into pieces and so putrefied. And about Estamenus in India there are abundance of Scorpions generated, only by corrupt rain-water standing in that place. And when one had planted the herb *Basilisca* [probably basil] on a wall in the room or place thereof he found two Scorpions. And some say that if a man chaw in his mouth fasting this herb basil, before he wash, and afterward lay the same abroad uncovered where no sun cometh at it for the space of seven nights, taking it in all the day-time, he shall at length find it transmitted into a Scorpion, with a tail of seven knots. Out of an herb *Sissumbria* putrefied, Scorpions are engendered. And out of the crocodile's eggs do many times come Scorpions, which at their first egression do kill their dam that hatched them. The Lybians, who among other nations are most of all troubled with Scorpions, do use to set their beds far from any wall, and very high also from the floor, and they also set the feet of their beds in vessels of water. Then the Scorpions in their hatred to mankind climb up to the ceiling, and one of them taketh hold upon that place in the house or ceiling over the bed wherein they find the man asleep, and so hangeth thereby, putting out and stretching his sting to hurt him, but finding it too short, and not being able to reach him, he suffereth another of his fellows to come and hang as fast by him as he doth upon his hold, and so that second giveth the wound,—and if that second be not able likewise, because of the distance, to come at the man, then they both admit a third to hang upon them, and so a fourth upon the third, and a fifth upon the fourth, until they have made themselves like a chain, to descend from the top to the bed wherein the man sleepeth, and the last striketh him; after which stroke he first of all runneth away by the back of his fellow, and every one again in order, till all of them have withdrawn themselves. It is thought that hares are never molested by Scorpions, because if a man or beast be anointed with the rennet of a hare, there is no Scorpion or spider will hurt him. Wild goats are also said to live without fear of Scorpions. The seed of nose-wort burnt or scorched doth drive away serpents, and resist Scorpions, and so doth the seed of violets and of wild parsnip. The smell of garlic and wild mints set on fire or strewed on the ground, and dittany have the same operations; and above all other, one of these Scorpions burned driveth away all his fellows which are within the smell thereof. By touching of hen-bane they lie dead and overcome, but if one touch them again with white Hellebore they revive. The sea-crab with basil in her mouth destroyeth the Scorpion, and so doth mushroom of trees. To conclude, the spittle of a man is death unto Scorpions.

Topsell, "Four-footed Beasts," pp. 750-57.

YOUR tongues like Scorpions
Both heal and poison.

Beaumont and Fletcher, "Philaster," iii. 1.

THEY which are stung with the Scorpion cannot be recovered but by the Scorpion. "Euphues' Golden Legacie."

Screech-owl.

TROILUS AND CRESSIDA, v. 10, 16.

AMONG diviners with crying he tokeneth adversity; and if he be still, he tokeneth prosperity.

Bartholomew (*Berthelet*), bk. xii. § 36.

Sea-maid.

MIDSUMMER NIGHT'S DREAM, ii. 1, 154.

V. Siren.

Sea-mew.

TEMPEST, ii. 2, 176.

[Sea-mew, a small gull.]

IN the Lake of Como, certain fishers in the winter did draw with their nets to the dry land a great sort of Sea-mews, seeming to be dead, which were joined together with their bills or nebs in one another's tail ; and being warmed with their guts, were found alive.

Lupton, "Notable Things," bk. vi. § 88.

Sedge.

TAMING OF THE SHREW, Induction, 2, 53.

SEDGE is an herb most hard and sharp, and hurteth never man but he toucheth it.

Bartholomew (*Berthelet*), bk. xvii. § 35.

SEDGES or sheargrass, whereof is made mats and hassocks to sit and kneel upon ; with the said Sedges is made Hamboroughs [*i.e.*, collars] for the necks of horses, instead of leather harness, and for other cartage and plough.

Batman's addition to *Bartholomew*, *loc. cit.*

Serpent.

ALL kind of Serpents and adders, that by kind may wrap and fold his own body, hath many corners and angles in such folding, and goeth never straight. And of adders is many manner kind ; and how many kind, so many manner venom, and how many species, so many manner malice, and so many manner sores and aches, as there are colours. And an adder grieveth most now with biting, now with blowing, now with smiting with the tail, and now with stinging, now with looking and sight. The Serpent *Dipsas* is so little, that he uneath [scarcely] is seen when men tread thereon, and the venom thereof slayeth ere it be felt, and he that dieth by that venom feeleth no sore. Some have two heads, as the adder *Amphisbæna*, one in the one end, and another in the tother end, and runneth and glideth and wriggleth with wrinkles, circles and draughts of the body after either head, as though one mouth were too little to cast venom. Also some Serpents have many heads; for some be doubled, and some trebled, and some quadrupled. And *Hydra* is a Serpent with many heads, and it is said that if one head be smitten off, three grow again—but this is a fable. The Serpent *Scytalis* shineth with diversity of speckles, that all that looketh thereon for wonder of the speckles hath liking to look thereon, and, for he is most slow in creeping, by a wonder of his diversity of speckles, he catcheth them that he may not follow in going and in creeping. And the Serpent *Enhydris* is a water-adder, and whoso is smitten of that adder, he swelleth into dropsy, and the dirt of an ox is remedy therefor. Also *Natrix* is an adder, and infecteth with venom each well that he cometh nigh. And some Serpents and adders lie in await for them that sleep, and if they find the mouth open of them or of other beasts, then they creep in, but against such adders a little beast fighteth as it were a little eft (and some men mean that it is a lizard), for he leapeth upon his face that sleepeth, and scratcheth with his feet to wake him and to warn him of the Serpent. And this little eft, when he waxeth old, his eyes waxeth blind, and then he goeth into an hole of a wall against the East, and openeth his eyes afterward when the sun is risen, and then his eyes healeth and taketh sight. And some manner Serpent dwell in the fire as it fareth of the salamander [*q.v.*]. Also some Serpents go forth and hold up the body from the breast upward, as the water-adder doth that hight *Chelydros*, and he infecteth the place that he glideth in, and maketh the sight smoky; and this Serpent beareth up the head, for if he bendeth while he runneth, he breaketh anon. And some be so swift and light of moving, that it seemeth that they fly, as the Serpent that hight *Jaculus* flieth as a dart, and leapeth into trees, and if he meeteth with any beast, he throweth himself thereupon, and slayeth it. Also in Arabia be Serpents called Sirens among many men; and they run swifter than horses, and therefore it is said that they fly, and their venom is so strong that death cometh tofore [before] biting, and tofore ache also. And the horned Serpent *Cerastes* hideth himself in gravel and sand, and sheweth his horns above to comfort beasts and fowls to come as it were to meat by shewing of horns; and hath horns like ram's horns, and beasts and fowls come thereto, and ween to find there a ram, and find a venomous Serpent when they have assayed. Also Boa is a Serpent full great in quantity, and is in Italy, and followeth flocks of neat and of bugles [buffaloes], and setteth himself guilefully to the udders of the beasts that be full of milk, and sucketh and slayeth them. The head of a Serpent scapeth and liveth, if it may scape with two fingers of the body, and therefore they put forth all the body for defence of the head. Also all Serpents have dim sight, and look awayward, and no wonder, for their eyes be not in the forehead

but in the temples, so that they may rather hear than see. Also no beast moveth the tongue so swiftly as the Serpent, for it moveth the tongue so swiftly, that it seemeth that it hath three tongues, yet it hath but one. Also the bodies of serpents be moist, so that where they glide and go, they infect the way, and mark it with a manner glymy [viscous] humour. Also Serpents live long and without meat. And they live so long time, that they put away their old skins, and become young again. The manner of changing of Serpents' skins seemeth wonderful enough; for the adder feeleth himself grieved with evil, or with age, and abstaineth and fasteth many days, that his skin may so the easilier be departed from the flesh; and then he tasteth a certain bitter herb, that maketh him vomit and cast, and so he casteth out the venomous humour that was cause of his sickness and his default, and batheth himself at the last, and moisteneth himself in water to temper and to nesh [soften] the tender skin. And so he seeketh a strait cliff [cleft] of a stone, or some strait den or some other thing, and entereth into a strait chine or den, and passeth through with a manner violence, and unlooseth himself cleanly of the old skin, and then he layeth himself in the sun, and drieth himself, and recovereth a new skin about the flesh, and taketh might and strength, and seeth more clear, and glideth and passeth and creepeth more strongly, and eateth more savourly than he did tofore the changing of the skin. Of the marrow of the ridge-bone [spine] of a dead man, a Serpent is gendered. And also it is said that a Serpent dreadeth a naked man, and dare not touch him, though he leap on him, when he is unclothed. And a fasting man's spittle is venom to Serpents, and Serpents die if they taste thereof. In winter-time Serpents lurk in darkness and dens, and their sight dimmeth for long abiding in darkness; then when they come out first of their dens in springing-time, they feel dimness of sight, and seek fennel, or the roots thereof and eat it, and doth away blindness. And the snail is not beguiled of remedy, nor the tortoise when they have eaten a Serpent's guts, for as they take heed that the venom creepeth and worketh, they seek wild marjoram, and find by taste thereof medicine against the venom of the Serpent. The very Serpent drinketh but little, and hateth the smell of rue, and fleëth therefore the weasel, when he hath eaten rue, and may not well flee when he smelleth rue. And a Serpent hath thirty ribs by the number of the days of a month. And Serpents fare as swallows' birds [do], for if their eyes be put out, yet their sight cometh again; and the tail of a Serpent groweth again if it be cut off, as the tail of a newt. Also the weasel fighteth against Serpents, and armeth himself with eating of rue, and fighteth namely against Serpents that eat mice, for the weasel hunteth and eateth mice. Also Serpents love well wine, and be therefore hunted with wine. And also a Serpent loveth passing well milk, and followeth the savour thereof, and therefore if a Serpent be crept into a man's womb, he may be drawn out with odour and smell of milk.

Bartholomew (*Berthelet*), bk. xviii. § 8.

If a water-snake be tied by the tail with a cord, and hanged up, and a vessel full of water set under the said snake,—after a certain time he will avoid out of his mouth a stone, which stone being taken out of the vessel, he drinks up all the water. Let this stone be tied to the belly of them that have the dropsy, and the water will be exhausted or drunk up, and it fully and wholly helps the party that hath the said dropsy.

Lupton, "Notable Things," bk. i. § 11.

If any do sprinkle his head with the powder of the skin that a snake doth cast off, gotten or gathered when the moon is in the full, being also in the first part of Aries, the Ram, he shall see terrible and fearful dreams. And if he shall have it under the plant of his foot, he shall be acceptable before magistrates and princes. *Ibid.*, bk. iv. § 54.

A certain country-man did sleep open-mouthed in the fields, a Serpent crept in at his mouth, and so into his body; but after the same man cured himself thereof with eating of garlic. But he infected his wife with poison, whereof she died, which was very rare and strange.

Ibid., bk. v. § 95.

A Serpent doth so hate the Ash-tree, that she will not come nigh the shadow of them, and therefore she goes far from them both morning and evening, because then they give the longest shadows, etc. *Ibid.*, bk. ix. § 8.

V. **Ash-tree.**

Serpents being within a circle made of Betony, they cannot go out of the same, but rather will die with beating themselves. *Ibid.*, bk. ix. § 28.

In Egypt as frogs and mice are engendered by showers of rain, so also are Serpents; and the longest hairs of women are easily turned into Serpents; and dung, being laid in a hollow place, subject to receive moisture engendereth Serpents.

Topsell, "History of Serpents," pp. 595-96.

[Topsell gives many other curious facts about Serpents, but as his treatise covers nearly forty folio pages, this sample must suffice.]

YOU have ate a snake, and are grown young, gamesome and rampant.

Beaumont and Fletcher, "The Elder Brothers," iv. 4.

HE hath left off o' late to feed on snakes;
His beard's turn'd white again.

Massinger, etc., "The Old Law," v. 1.

YOUR viper wine
So much in practice with grey-bearded gallants,
But vappa to the nectar of her lips.

Ibid., "Believe as You List," iv. 1.

THAT men may appear to be headless:—Take the slough of a snake, and auripigment [arsenic] and Greek pitch, and the wax of young bees, and ass's blood, and pound them all, and put them in a rough jar full of water, and make it boil on a slow fire, and then let it cool, and make a taper of it, and every man who shall be illuminated by that taper will seem to be headless.

Albertus Magnus, "Of the Wonders of the World."

IF you wish to kill a Serpent quickly, take as much as you please of *Aristolochia Rotunda*, and pound it well, and take a frog of the woods or of the fields, and pound and mix it with the *Aristolochia*, and put with it something burnt, and write with it on a paper, or anything that you prefer, and throw it to the Serpents. *Ibid.*

THAT a house may seem quite green and full of Serpents and fearful images, take the skin of a Serpent, and the blood of another male Serpent, and the fat of another Serpent, collect all these three things, and put them in a cere-cloth, and kindle it in a new lamp. *Ibid.*

NOTE that if you boil a Serpent or a worm, and give of the fat of that worm to any man to eat, he will understand when they sing. [He does not explain who are "they," but from the previous article "they" may be mice or Serpents.] This has been proved.

Albertus Magnus, "Of the Wonders of the World."

IN a garden of the suburbs [of Aleppo] I did see a Serpent of wonderful bigness, and they report that the male Serpent and young ones being killed by certain boys, this she-serpent, observing the water where the boys used to drink, did poison the same, so as many of the boys died thereof; and that the citizens thereupon came out to kill her, but seeing her lie with her face upward, as complaining to the heavens that her revenge was just, that they, touched with a superstitious conceit, let her alone; finally that this Serpent had lived here many ages, and was of incredible years. *Fynes Moryson*, "Itinerary," part i. p. 246.

IF a snake did bite a Cappadocian, the man's blood was poison to the snake, and killed him.

Purchas' "Pilgrims," p. 320 (ed. 1616).

SERPENTS are plentifully engendered of much rain, or effusion of men's blood in war. *Ibid.*, p. 560.

THE cucurijuba is a fresh-water snake [in Brazil], toothed like a dog; it catcheth a man, cow, stag, or other prey, winding it with the tail, and so swalloweth it whole; after which she lies and rots, the ravens and crows eating her all but the bones, to which after groweth new flesh, by life derived from the head, which is hidden all this while in the mire. *Ibid.*, p. 843.

Sheep.

A SHEEP is a nesh [soft] beast, and beareth wool, and is unarmed in body, and pleasing in heart. And if Sheep conceive toward the Northern wind, they conceive males; and if they conceive toward the Southern wind, then they conceive females. And such as the veins be under the Sheep's tongue, of such colour is the lamb when he is eaned. And cold water of the North is good to them in summer, and warm water of the South is good to them in harvest. And herds know which of them may dure [endure] in winter; for upon some is found ice, and upon some none ice is found; and some of them be feeble, and may not shake off the ice; and those that have long tails may worse away-with winter than those that have broad tails. And wool of Sheep that a wolf eateth is infected; and the cloth that is made thereof is lousy. Also in Sheep is less wit and understanding than in another four-footed

beast. Also thundering maketh solitary Sheep to cast their lambs; the remedy and help thereof is to gather and bring them together into one flock.

Bartholomew (*Berthelet*), bk. xviii. § 81.

Of Sheep, their wool is a singular benefit in a commonwealth, especially the Cotswold wool for fineness. And in Bartholomew's time, the staple for wool was not so well husbanded as it hath been since. The increase of pasture for Sheep hath so much decreased the tillage of corn, that, until it be restored again, there will grow a poor commonwealth. The more Sheep, the dearer [cheaper?] the wool, the flesh and the fell; the more Sheep, the dearer corn and grain, beside beef, butter, eggs and cheese. Pastures consume tillage; the want of tillage breeds beggars, decays villages, hamlets and upland towns. It is better to want wool than corn, Sheep than men, but excess and prodigality, which cannot away-with measure, have brought this England to great penury.

Batman's addition to *Bartholomew*, bk. xviii. § 81.

Sheep are wont to follow them that stop their ears with their wool. *Lupton*, "Notable Things," bk. v. § 48.

About Erythrea, there is such abundance of good pasture and herbs so grateful to Sheep, that if they be not let blood once in thirty days, they perish by suffocation, and the milk of those Sheep yieldeth no whey. The rams of England have greater horns than any other rams in the world, and sometimes they have four or six horns on their head, as hath been often seen. In very cold countries, when snow and winter covereth the earth, then Sheep have no galls, but in the summer when they go abroad again to feed in the fields, they are replenished with galls. Sheep, when they have eaten *Eryngium* [sea-holly], all stand still, and have no power to go out of their pastures till their keeper come and take it out of their mouths. The Sheep of Lydia and Macedonia grow fat with eating of fishes. If there appear upon grass spiders' webs, or cobwebs which bear up little drops of water, then they must not be suffered to feed in those places for fear of poisoning. Because the head of Sheep is most weak, therefore it ought to be fed turned from the sun.

Topsell, "Four-footed Beasts," pp. 464-69.

Shell.

King Lear, i. 5, 26.

It is an usual thing to crush and break both egg- and fish-shells, so soon as ever the meat is supped and eaten out of them, or else to bore the same through with a spoon, steel or bodkin. [Marginal note: Because afterwards no witches might prick them with a needle in the name and behalf of those whom they would hurt and mischief, according to the practice of pricking the images of any person in wax, used in the witchcraft of these days.]

Holland's Pliny, bk. xxviii. ch. iii. and note.

Sherris.

ii. KING HENRY IV., iv. 3, 110, etc.

[Wine of Xeres.]

V. Sack.

Shough.

MACBETH, iii. 1, 94.

[A kind of rough-haired dog.]

Shrew[-mouse].

[*Shrew* is only used by Shakespeare of woman. The word has the same meaning (*v. infra*).]

A SHREW-MOUSE quasi shrewd mouse, which by biting cattle so venometh them that they die, whereof came our English "I beshrew thee," when we wish ill.

Minsheu's Dictionary, *s.v.*

IT is a ravening beast, feigning itself gentle and tame, but being touched, it biteth deep, and poisoneth deadly. They annoy vines, and are seldom taken, except in cold; thay frequent ox-dung. If they fall into a cart-road, they die and cannot get forth again. They go very slowly, they are fraudulent, and take their prey by deceit. Many times they gnaw the ox's hoofs in the stable. They love the rotten flesh of a raven. The Shrew being cut and applied in the manner of a plaister doth effectually cure her own bites. The dust of a cart-rut [in which a Shrew has died] being taken and sprinkled into the wounds made by her poisonous teeth is a very excellent and present remedy for the curing of the same. If horses or any other labouring creature do feed in that pasture or grass in which a Shrew shall put forth her venom or poison in, they will presently die.

Topsell, "Four-footed Beasts," pp. 415-20.

To keep beasts safe that the blind mouse called a Shrew do not bite them: Enclose the same mouse quick in chalk, which when it is hard, hang the same about the neck of the beast that you would keep safe from biting; and it is most certain, that he shall not be touched nor bitten. *Lupton*, "Notable Things," bk. vii. § 52.

If a Shrew, I take it to be the blind mouse, doth chance to go over any part of any beast, that part of the beast will after be lame. This I know to be true.

Ibid., bk. x. § 11.

Shrimp.

ii. King Henry VI., ii. 3, 23.

[Addressed to a salmon] one
That for the calmest and fresh time o' the year
Dost live in shallow rivers, rank'st thyself
With silly smelts and Shrimps.

Webster, "Duchess of Malfi," iii. 5.

You shall eat nothing but Shrimp porridge for a fortnight. *Brome*, "Sparagus Garden," ii. 3.

Silk.

So often as I consider that some ten thousands of Silkworms, labouring continually night and day, can hardly make three ounces of silk,—so often do I condemn the excessive profusion and luxuriousness of men in such costly things, who defile with dirt silks and velvets, that were formerly the ornaments of kings, and make no more reckoning of them now than of an old tattered cloak, as if they were ashamed to esteem better of an honourable thing than of a base, and were wholly bent upon waste. Amongst the English a silken habit is so much loved and valued, that they despise their own wool, which compared with silk is not contemptible, and is the most profitable and the greatest merchandise of the kingdom. But time will make them forego this wantonness, when they shall observe that their moneys are treasured up in Italy at that time, when they stand in need of it for their public or private affairs.

Dr. Thos. Mouffet, "Theatre of Insects," p. 1033.

Siren.

COMEDY OF ERRORS, iii. 2, 47.

THE mermaiden hight Siren, and is a sea-beast wonderly shapen, and draweth shipmen to peril by sweetness of song. And some men say that they are fishes of the sea in likeness of women. Sirens be great dragons flying with crests, as some men trow. And some men feign that there are three Sirens somedeal maidens, and somedeal fowls with

claws and wings, but the sooth is, that they were strong whores, that drew men that passed by them to poverty and to mischief. And in Arabia be serpents with wings, that be called Sirens, and run more swiftly than horses, and do fly with wings, and their venom is so strong that death is felt sooner than ache or sore. And Siren is a beast of the sea, wonderly shapen as a maid from the navel upward, and a fish from the navel downward, and this wonderful beast is glad and merry in tempest, and sad and heavy in fair weather. With sweetness of song this beast maketh shipmen to sleep, and when she seeth that they be asleep, she goeth into the ship, and ravisheth which she may take with her, and bringeth him into a dry place, and [the rest is indecent]. *Bartholomew* (*Berthelet*), bk. xviii. § 97.

Its face is horrible, its hair very long and filthy. And it appears with its young which it carries in its arms; and when sailors see it, they are much afeared, and throw it an empty bottle, with which it plays, until the ship has passed by.

Hortus Sanitatis, bk. iv. § 83.

Slug.

Comedy of Errors, ii. 2, 196.

V. Snail.

Snail.

As You Like It, iv. 1, 54.

Snail is a worm of slime, and breedeth of slime, and is therefore alway foul and unclean; and is a manner snake, and is an horned worm. And such worms be gendered principally in corrupt air and rain.

Bartholomew (*Berthelet*), bk. xviii. § 70.

Neither have I
Dress'd Snails or mushrooms curiously before him.

Ben Jonson, "Every Man in his Humour," ii. 5.

Some men trow, though it be not be believed, that the ship goeth slower, if he beareth the right foot of a Snail [*Batman* translates "*testudo*" here by "tortoise" instead of "Snail"].

Bartholomew (*Berthelet*), bk. xviii. § 107.

Snails without their shells, or otherwise with their shells stamped and mixed sometimes with cheese-lope or rennet, do draw out thorns, or any other thing out of the flesh, though never so deep, if they be applied to the place.

Lupton "Notable Things," bk. 1 § 100.

The two horns of a snail borne upon a man will pluck away carnal or fleshly lust from the bearer thereof.

Ibid., bk. ix. § 17.

Snake. *V.* Serpent.

Sore, Sorel.

LOVE'S LABOUR'S LOST, iv. 2, 59, 60.

Sore, a deer of four years old.
Sorel, a deer of three years. *Minsheu's* Dictionary, *s.v.*

Sow.

MACBETH, iv. 1, 64.

A sow rooteth and diggeth the earth to get her meat and food, and overturneth and rooteth that she may come with the teeth to mores [roots] and roots. And the young sow conceiveth against the evenness of day and night in springing-time, and farroweth sometime twenty pigs at once, but she eateth all sometime, out-taken the first, for he is most kindly to her, and she giveth him alway the first teat. The Sow is an unclean beast, and a right great glutton, and coveteth and desireth baths, fens and puddles, and resteth herself therein, and waxeth fat. And the seventh part of her meat turneth into hair and blood, and into other such. *Bartholomew* (*Berthelet*), bk. xviii. § 99.

V. Swine, Boar.

Spaniel.

MIDSUMMER NIGHT'S DREAM, ii. 1, 203-7.

THE best sort of these dogs came from Spain.

Minsheu's Dictionary, *s.v.*

THE water-spagnel is taught by his master to seek for things that are lost (by words and tokens), and if he meet any person that hath taken them up, he ceaseth not to bay at him, and follow him, till he appear in his master's presence. They use to shear their hinder parts, that so they may be the less annoyed in swimming.

I may here also add the land-spagnel attending a hawk who are taught by falconers to retrieve and raise partridges. They are for the most part white, or spotted with red or black. *Topsell*, "Four-footed Beasts," p. 122.

Sparrow.

TROILUS AND CRESSIDA, ii. 1, 77.
KING JOHN, i. 1, 231.

THE Sparrow is an unsteadfast bird with voice and jangling, and is a full hot bird and lecherous, and the flesh of them oft taken in meat exciteth to carnal lust. Sparrows lay many eggs, and are full busy to bring up their birds, and to feed them. And she keepeth her nest clean without dirt, and therefore she throweth the dirt of her birds out of the nest, and compelleth her birds to throw their dirt out of the nest; and they feed their birds with attercops, worms and flies; and they eat venomous seeds, as of henbane without hurt; and they have sometime leprosy and the falling-evil. And the Sparrow dreadeth the weasel, and hateth her, and crieth and warneth if the weasel cometh. And waileth, and biteth, and billeth for to have the nests of swallows. And birds [*i.e.*, young birds], that other Sparrows leave by some hap, they gather and feed and nourish, as they were their own. And if it happeth that one of them is taken in a gin, or in other manner of wise, she crieth for help—and a multitude of Sparrows be gathered together to deliver that that is taken, and speed and haste with all their might.

Bartholomew (*Berthelet*), bk. xii. § 32.

Merchant's Wife: What's your cock-sparrows a dozen?
Seller: A penny, mistress.

"Histriomastrix," ii. 1, 77.

[But this was during the reign of Plenty, when corn was 2s. 6d. a quarter.

Sparrows, especially cock-sparrows, as aphrodisiacs were a constant ingredient of cullises; so were Sparrows' eggs.]

IF any will make their hands white, let them mix the dung of Sparrows in warm water, and wash them therewith; or let them seethe the roots of nettles in that water, and therewith wash their hands.

Lupton, "Notable Things," bk. ii. § 69.

It is said that no Sparrows have ever been seen at a place called Lindham in the moors below Hatfield [Yorkshire].

Camden's "Britannia," col. 850.

Spawn.

CORIOLANUS, ii. 2, 82.

This is to be noted, that the foresaid engendering of [fish] is not sufficient to accomplish generation, unless, when their eggs be laid or spawn cast, both male and female take it between them, and keep a-turning of it, thereby to breathe a lively spirit into it, and, as it were, besprinkle it with a vital dew, as it floateth upon the water. But turn they it and toss it, breathe they upon it as much as they will, yet all those little eggs of their spawn do not hit and come to proof; for if they did, all seas and lakes, and all rivers and pools, would be so pestered full with fishes, that a man would see nothing else.

Holland's Pliny, bk. ix. ch. l.

V. Fish.

Spear-grass.

i. KING HENRY IV., ii. 4, 340.

SPEAR-GRASS is good for the sciatica, or the gout.

Lupton, "Notable Things," bk. ii. § 91.

TOUCHING the grass, which, by reason of the pricks that it bears is named *Aculeatum*, there be three sorts of it; the first is that which ordinarily hath five such pricks in the head or top thereof, and thereupon they call it Pentadactylon, the Five-finger grass; these pricks, when they be wound together, they use to put up into the nostrils, and draw them down again, for to make the nose bleed.

Holland's Pliny, bk. xxv. ch. xix.

Spermaceti.

Telling me the sovereign'st thing on earth
Was parmaceti for an inward bruise.

i. KING HENRY IV., i. 3. 58.

Ticket: I bruised my side e'en now against a form's edge.
Rufflit: Parmaceti, Sir, is very good, or the fresh skin of a flayed cat.

Brome, "The City Wit," Act. v.

My dear mummia, my balsamum, my Spermaceti!

Ben Jonson, "Poetaster," ii. 1.

Spice.

Much Spice is a thief, so is candle and fire,
Sweet sauce is as crafty, as ever was friar.

Tusser, "Five Hundred Points." Afternoon Works, § 14.

[Spice is a thief, because it was bought, and not home-grown, and also because it increases appetite.]

V. **Cloves, Nutmeg,** etc.

Spider.

WINTER'S TALE, ii. 1, 39.

The venomous Spinner is a little creeping beast with many feet, and hath vi. feet or viii., and hath alway feet even and not odd; and that is very needful, that his going and passing be alway even, as the charge is and burthens. In the end of springing-time, and in the beginning of summer, and sometime in harvest, and in the beginning of winter, Spinners be most grievous, and their biting most venomous. And a manner kind of Spinners hunteth a little eft, and when they find him, they begin to weave upon him, and all about, for to bind strongly his mouth, and leap then upon him, and sting him till he dieth. Wonder it is, how the matter of threads that come of the womb of the Spinner may endure so great a work, and weaving of so great a web. Also in Spinners be tokens of divination, and of knowing what weather shall fall,—for oft by weathers that shall fall, some spin and weave higher or lower. Also multitude of Spinners is token of much rain. Also sometime Spinners weave and make webs about burgeoning and buds of vines, and also about flowers and blossoms of trees, and by such beclipping [embracing] of such cobwebs, both trees and vines be lost where they burgeon and bloom. The biting of the Spinner that hight *sphalangio* is venomous and slayeth, but there be remedy and succour the sooner; but the virtue of plantain slayeth the venom thereof, if it be laid thereto in due manner, and therefore other worms, as efts and frogs, that dread the stinging of Spinners, defend themselves with juice of plantain. And though the Spinner be venomous, yet the web that cometh out of the guts thereof is not venomous, but is accounted full good and profitable to the use of medicine. And a manner Spinner hight *spalana*, and is like to an ant, but he is much more of body, and hath a red head, and the other deal of the body is black, sprung [sprinkled] with white specks; and his smiting is more bitter and more sore than the biting of the serpent Viper; and this Spinner liveth most nigh furnaces, ovens and mills; and the remedy against its biting or smiting is to shew to him that is bitten or smitten another Spinner of the same kind; and be therefore kept when they be found dead; the skin thereof stamped and drunk is medicine against biting of the weasel. Also another Spinner is rough with a great head, and the soreness and ache of his stinging is as it were the ache and soreness of a scorpion, and by his biting the knees shake and faileth, and also of the biting cometh blindness and spewing. And another manner Spinner is like to an ant with a great head, and hath a black body with white specks. His biting paineth and acheth as stinging of wasps, and hight *formicaleon* [ant-lion] for he hunteth ants, but sparrows and other fowls devour him, as they do ants. Against all biting of Spinners, the remedy is the brain of a capon drunk in sweet wine with a little pepper; also flies stamped and laid to the biting draweth out the venom, and abateth the ache and sore. And the same doth ashes of a ram's claw [*Bartholomew*, "lamb's rennet"] with honey.

Bartholomew (*Berthelet*), bk. xviii. § 11.

THE Spider is a worm of the air.

Hortus Sanitatis, bk. ii. § 11.

THE great number of Spiders do foreshow that the summer following will be pestiferous and plaguy.

Lupton, "Notable Things," bk. ii. § 82.

WHEN houses are ready to drop down, Spiders with their cobwebs first of all fall, and get them away packing, alter their climate to some other surer place and dwelling to rest in. The Spider beareth a deadly feud and mortal hatred to serpents; for if so be the serpent at any time lie in the shadow under any tree to cool himself, where Spiders do resort, some one of them levelleth directly at him, and with such a violence striketh and dasheth at his head with her beak or snout, that her enemy withal making a whizzing noise, and being driven into a giddiness, turning round, hisseth, being neither able to break asunder the thread that cometh from above, nor yet hath force enough to escape it [and so the snake is killed].

Topsell, "History of Serpents," pp. 782-83.

THE Spider feedeth of the corruption that she findeth in the flowers and fruits that are in the gardens, whereas the bee gathereth her honey out of the best and fairest flower she can find.

"History of Hamblet, Prince of Denmark."

THE poets' Arachne doth never weave her entangling web near the cypress-tree.

Walkington, "Optic Glass of Humours," p. 96.

Spinner.

ROMEO AND JULIET, i. 4, 59.

V. Spider.

Sponge.

HAMLET iv. 2, 22.

WHEN a Sponge is thrown into wine mixed with water, then taken out and squeezed, the water comes out of it, and the wine remains, and if it be not mixed, nothing comes out.

Albertus Magnus, "Of the Wonders of the World."

Sprat.

ALL'S WELL THAT ENDS WELL, iii. 6, 112.

GREAT lords sometimes
For change leave calver'd salmon, and eat Sprats.
Massinger, "Guardian," iv. 2

ALL-SAINTS do lay for pork and souse
For Sprats and spurlings for their house.
Tusser, "The Farmer's Daily Diet."

[Sprats were caught at the mouth of the Thames, but young herrings were frequently substituted for them. The peck of Sprats or young herrings were sometimes sold at Billingsgate for two-pence. The best Sprats came from Orfordness, and Dunwich-bay. From "England's Way to Win Wealth" (1614)].

BROILED red Sprat.
Middleton, "Blurt, Master Constable," iii. 3, 205.

Stag. *V.* Hart.

Stockfish.

TEMPEST, iii. 2, 79.

ABOUT the Isle Ebusus, the Stock-fish is much called for; whereas in other places it is counted but a base, muddy and filthy fish. *Holland's Pliny*, bk. ix. ch. xviii.

[From this it is clear that the Stockfish was a distinct species, but the word usually means dried fish as distinct from fresh.]

THE Stock-fish-mongers [are seated] in Thames Street; wet-fish-mongers in Knightriders Street and Bridge Street. Stock-fishes, so called for dried fishes of all sorts, as lings, haberdines, and other. *Stow's* "Survey."

[Stephano, in the passage quoted from the "Tempest," means that he will beat Trinculo; so *Middleton*, "Blurt, Master Constable," iii. 3, 17: "I do not love to handle these dried Stockfishes, that ask so much tawing"; and *Webster*, "Westward Ho!" v. 4: "Have you Stockfish in hand that you beat so hard?" and *Beaumont and Fletcher*, "The Captain," iii. 3: "Beat him soft like Stockfish."]

[Stork.]

[The word is not actually used by Shakespeare, but the account of the bird is interesting.]

A Stork is a water-fowl, and purgeth herself with her own bill; for when she feeleth herself grieved with much meat, she taketh sea-water in her bill, and putteth it in at her hinder hole, and so into her guts. Also this bird eateth eggs of adders and serpents, and beareth them for best meat to her birds. And they leave not lightly their first nest, except they be compelled. But ere they go into other countries against winter, they fill their nests with

earth, and draw the twigs and thorns of their nests with fen, that no tempest of wind should break it nor throw it down in winter. While the female liveth the male keepeth truly to her in nest. And if the male espy in any wise, that the female hath broke spousehead, she shall no more dwell with him, but he beateth and striketh her with his bill, and slayeth her if he may. Storks fly over the sea in flocks, and in their passing crows fly with them, and pass tofore them, as it were leading the Storks, and withstand with all their might fowls that hate Storks.

Bartholomew (*Berthelet*), bk. xii. § 8.

STORKS nourish their parents when oppressed with age.

Minsheu's Dictionary, *s.v.*

Stover.

TEMPEST, iv. 1, 63.

Stover, or *Estover*—Fodder.

Minsheu's Dictionary, *s.v.*

THRESH barley as yet but as need shall require
Fresh threshed for Stover thy cattle desire.

Tusser, "November's Husbandry."

Straw.

SOME Straw is kept to fodder of beasts, for it is first meat that is laid tofore beasts, namely in some countries, as in Tuscany. And the kind thereof is cold that it suffereth not snow that falleth to shed [melt], and is so hot that it compelleth apples for to ripen.

Bartholomew (*Berthelet*), bk. xvii. § 65.

Strawberry.

The strawberry grows underneath the nettle,
And wholesome berries thrive and ripen best
Neighbour'd by fruit of baser quality.

KING HENRY V., i. 1, 60.

My lord of Ely,
When I was last in Holborn,
I saw good strawberries in your garden there.

KING RICHARD III., iii. 4, 34.

STRAWBERRIES do grow upon hills and vallies, likewise in woods, and other such places that be somewhat shadowy; they prosper well in gardens. *Gerard's* "Herbal," *s.v.*

[Evelyn ("Kalendarium Hortense") enumerates the following kinds: Common Wood, English Garden, American or Virginian, Polonian, White Coped, Long Red, Green, Scarlet, etc.]

MANY have been helped that have had foul and leprous faces, only with the washing the same with distilled water of Strawberries; the Strawberries first put into a close glass, and so putrified in horse-dung.

Lupton's "Notable Things," bk. iii. § 82.

[Strawberries were eaten with cream and sugar, or with claret and sugar, in the Continental fashion, according to Dr. Hart ("Diet of the Diseased," p. 60), where he orders them to be taken before other food.]

Sugar.

i. KING HENRY IV. ii., 4, 25.
LOVE'S LABOUR'S LOST, v. 2, 231.

[The finest Sugar came from Barbary (*Beaumont and Fletcher's* "Beggar's Bush," iv. 3, "*Webster's* "Northward Ho!" ii. 1, and *Marston's* "What You Will," ii. 2, 83). Parmesan Sugar was also a costly luxury (*Chapman*, "The Ball," Act iii.). The practice of drinking Sugar with wine was exclusively English, according to Fynes Moryson. There are frequent allusions to it, *e.g.*:

Fill us of your nippitate, sir,
But hear ye, boy?
Bring Sugar in white paper, not in brown.

"Look about You," sc. 21.

A pound and a half of sugar cost in 1555 1s. 7½d. (*Brand's* "Popular Antiquities," vol. i. p. 425, *note.*)
Sugar will make a man kind (*Brand*, vol. ii. p. 95, *note*).]

Sugar-candy.

i. KING HENRY IV., iii. 3, 180.

HE's a mere stick of Sugar-candy
You may look quite through him.

Webster, "Duchess of Malfi," iii. 1.

Sugar-sop.

[Name of a servant in "Taming of the Shrew," iv. 1, 95.
A sweetmeat in *Beaumont and Fletcher's* "Monsieur Thomas," ii. 3.]

Sweet Marjoram. *V.* Marjoram.

Swine.

THE Swine froteth [rubs] and walloweth in dirt and in fen [mud], and diveth in slime, and bawdieth himself therewith, and resteth in a stinking place. And some Swine be tame, and some wild. And among the tame the males be called Boars and Barrows, and the females be called Sows; and they dig and root and seek meat under earth. A Swine dieth if he loseth an eye. And Swine have many sicknesses, and hold their heads aside; and lie more on the right side than on the left; and wax fat in forty days; and fat sooner, if they suffer hunger three days in the beginning of the feeding. Swine love each other, and know each other's voice, and therefore, if any cry, they cry all, and labour to help each other with all their might. Tame Swine grunt in going, and in lying, and in sleeping, and namely if they be right fat. And Swine sleep faster in May than in other times of the year, and that cometh of fumosity that stoppeth their brain that time. The male hath more teeth than the female. And when Swine be great, it doeth them good to eat berries, and also bathing in hot water delighteth them. And they be let blood on the vein under the tongue.

Bartholomew (*Berthelet*), bk. xviii. § 87.

Sycamore.

ROMEO AND JULIET, i. 1, 128.

SYCAMORE is a nice fig-tree, as it were a fool, and beareth certain sweet fruit that is never ripe at the fall.

Bartholomew (*Berthelet*), bk. xvii. § 148.

IT bringeth forth fruit three or four times in one year, and oftener if it be scraped with an iron knife, or other like instrument. We call it in English, Sycamore-tree, and also mulberry-fig-tree.

Gerard's "Herbal," *s.v.*

Tadpole.

KING LEAR, iii. 4, 135.

V. Frog.

Tassel-gentle.

ROMEO AND JULIET, ii. 2, 160.

TASSEL, or Tiercel, or the male of a hawk.
Minsheu's Dictionary, *s.v.*

LONG-WINGED Hawks, as the falcon gentle, and her Tiercel. *Markham's* "Husbandry" ("Of Hawks"), ch. i.

THEN for an evening flight
A Tiercel-gentle. *Massinger*, "Guardian," i. 1.

I SHOULD not be so fond to mistake a Jenny Howlet for a Tassel-gentle. *Brome*, "The Northern Lass," iii. 2.

[Malone quotes from an old treatise on hawking, name not given: "The names of all manner of hawks, and to whom they belong:—For a *Prince*. There is a falcon gentle, and a Tiercel gentle; and these are for a prince."]

Tench.

i. KING HENRY IV., ii. 1, 17, 18.

["Stung like a Tench" may perhaps refer to the small size of the scales of this fish. Nares quotes *Walton's* "Complete Angler" (part i. ch. xi.): "That the Tench is the physician of fishes, for the pike especially; and that the pike, being either sick or hurt, is cured by the touch of the Trench."]

I LONG to see this fish. I wonder whether
They will cut up his belly; they say a Tench
Will make him whole again.
Jasper Mayne, "The City Match," iii. 2 (1639).

V. **Fish.**

Thistle.

MUCH ADO ABOUT NOTHING, iii. 4, 76.

THISTLE is a manner herb or a weed with pricks; the kind thereof is biting and cruel, therefore the juice thereof cureth the falling of the hair. The root thereof sod in water giveth appetite to drinkers, and it is no wonder though women desire it, for it helpeth the conception of male children. *Barthölomew* (*Berthelet*), bk. xvii. § 36.

THE common Thistle, whereof the greatest quantity of down is gathered for divers purposes, as well by the poor to stop pillows, cushions, and beds for want of feathers, as also bought of the rich upholsters to mix with the feathers and down they do sell, which deceit would be looked into. The leaves and roots hereof are a remedy for those that have their bodies drawn backward.

Gerard's "Herbal," bk. ii. ch. ccccxxvi.

THE tender leaves of our Lady's Thistle, the prickles taken off, are sometimes used to be eaten with other herbs. The seeds being drunk are a remedy for infants that have their sinews drawn together, and for those that are bitten of serpents; and it is thought to drive away serpents, if it be but hanged about the neck.

Ibid., ch. ccccxxvii.

THE root of carline Thistle is an enemy to all manner of poisons; it doth not only drive away infections of the plague, but also cureth the same, if it be drunk in time. And it is given to those that have been dry-beaten, and fallen from some high place. *Ibid.*, ch. ccccxxxi.

V. Carduus Benedictus.

Thorn.

A THORN is a tree with sharp pricks, and is as it were armed with pricks against wrongs of them that touch it. And properly to speak, the thorn is the prick that groweth out of the thorn or of herbs and trees with pricks, and the prick springeth out of the stock or of the stalk, and is great next to the tree and stalk, and sharp outward at the point. And it is not the intent of kind that trees be sharp with pricks and thorns; but it happeth and cometh of unfastness and unsadness of the tree, by the which cold humour is drawn that is but little sodden, and is drawn and passeth by pores and holes outward, and is harded by heat of the sun, and made a thorn or a prick, and is made small and sharp at the end for scarcity of matter, and sometime is sharp and somedeal bending, as it fareth in Briers and Roseres [Rose-trees]; sometime the point is a-reared upright. Oft growing of thorns is token of barren land and untilled. And it is as it were a general rule, that all shrubs and trees with many thorns and pricks be wounden and wreathed together, and be clipped and succoured and defended each with other, and none of them hurteth other. And when they be felled or rooted up, they be bound in faggots and in heaps, and burnt in ovens and in furnaces, and for thorns be kindly dry, they be soon kindled in the fire, and give a strong light, and sparkleth, and cracketh, and maketh much noise, and soon after they be brought all to nought. Of thorns men make hedges and pavises [" large shields," according to Halliwell and Minsheu, but here certainly "fences"—*sepes*, *Bartholomew*], with which men defend and succour themselves and their own. *Bartholomew* (*Berthelet*), bk. xvii. § 149.

Throstle, Thrush.

MIDSUMMER NIGHT'S DREAM, iii. 1, 130.
WINTER'S TALE, iv. 3, 10.

Thrash, *throssel* or *mavis*. *Minsheu's* Dictionary, *s.v.*

THE Throstles or mavises all summer be painted about the neck with sundry colours, but in winter they be all of one colour. *Holland's Pliny*, bk. x. ch. xxix.

Thyme.

MIDSUMMER NIGHT'S DREAM, ii. 1, 249.

V. Cabbage.

Tick.

TROILUS AND CRESSIDA, iii. 2, 315.

TICK—a dog-louse. *Minsheu's* Dictionary, *s.v.*

THERE is a creature which hath evermore the head fast sticking within the skin of a beast, and so by sucking of blood liveth, and swells withal: the only living creature of all other that hath no way at all to rid excrements out of the body; by reason whereof when it is too full, the skin doth crack and burst, and so his very food is cause of his death. In kine and oxen they be common, and otherwhiles in dogs, who are pestered with these Ticks. And in sheep and goats, a man shall find none other but Ticks.

Holland's Pliny, bk. xi. ch. xxxiv.

[Topsell (*s.v.* "Sheep," p. 479) distinguishes between lice and Ticks of sheep.]

Tiercel, or Tercel.

TROILUS AND CRESSIDA, iii. 2, 56.

V. Tassel-gentle.

Tiger.

THE Tiger is the most swiftest beast in flight, as it were an arrow, and is a beast distingued with divers specks, and the river Tigris hath the name of this beast, for it is most swiftest of all floods. And the whelp is all glimy and sinewy. And the hunter taketh away the whelps, and fleëth soon away on the most swift horse that he may have, and when the wild beast cometh and findeth the den void, and the whelps away; then he riseth headlong, and followeth him by smell; and when the hunter heareth the grutching [grumbling, growling] of that beast that runneth after him, he throweth down one of the whelps, and the mother taketh the whelp in her mouth, and beareth him into her den, and layeth him therein, and cometh again after the hunter; but in the mean time the hunter taketh a ship, and hath with him the other whelps, and scapeth in that wise; and so her fierceness standeth in no stead; and the male recketh not of the whelps. And he that will bear away the whelps leaveth in the way great mirrors, and the mother followeth and findeth the mirrors in the way, and looketh on them, and seëth her own shadow and image therein, and weeneth that she seëth her children therein, and is long occupied therefore to deliver her children out of the glass, and so the hunter hath time and space for to escape. And in the more Hyrcania breedeth many beasts of this kind.

Bartholomew (*Berthelet*), bk. xviii. § 105.

A TIGER is bigger than the greatest horse. It hath been falsely believed that all Tigers be females, and that they engender with the wind. The male is seldom taken, because at the sight of a man he runneth away. When they hear the sound of bells and timbrels, they grow into such a rage and madness, that they tear their own flesh from their backs. The Indians near the River Ganges have a certain herb growing like Bugloss, which they take and press the juice out of it, and in still, silent, calm nights, they pour the same down at the mouth of the Tiger's den, by virtue whereof the Tigers are continually enclosed, not daring to come out over it through some secret opposition in nature, but famish and die, howling in their caves through intolerable hunger. The manner of this beast is, when she seeth that her young ones are shipped away, she maketh so great lamentation upon the sea-shore, howling, braying and ranking [perhaps "raging": Spenser uses the adverb "rank" in this sense], that many times she dieth in the same place; but if she recover all her young ones again, she departeth with unspeakable joy, without taking any revenge for their offered injury.

Topsell, "Four-footed Beasts," pp. 548-9..

THE Tiger as fierce and cruel as lions, making prey of man and beast, yet rather devouring black men than white; whose mustachios are holden for mortal poison, and, being given in meats, cause men to die mad.

Purchas' "Pilgrims," p. 559 (ed. 1616).

Tike, Tyke.

KING LEAR, iii. 6, 73.
KING HENRY V., ii. 1, 31.

[Tyke is now so well known a word for "cur," that Steevens' and Malone's notes on the latter passage seem to us ridiculous.]

Toad.

A TOAD is a manner venomous frog, and dwelleth both in water and in land; and he changeth his skin in age; and eateth alway certain herbs, and keepeth and holdeth alway venom, and fighteth against the common spinner [spider], and overcometh their venom and biting, by benefit of plantain; and his venom is accounted most cold, and [a]stonieth,—therefore each member that he toucheth, it maketh less feeling, as it were frore [frozen]; and is a venomous beast, and comforteth therefore himself at each touching. And the more he is touched, the more he swelleth; and as many specks as he hath under the womb, so many manner wise his venom is accounted grievous. And he hath eyes, as though they were fire, shining, and the worse he is, the more burning is his sight, and though he hath clear eyes, yet he hateth the light of the sun, and seeketh dark places, and fleëth to dens, when the sun riseth. This frog loveth sweet herbs, and eateth the roots of them; but in eating, he infecteth and corrupteth both roots and herbs; therefore oft in gardens is rue set, that is venom and enemy to Toads, and to other venomous worms, for by virtue of rue, they be chased away, and may not come to other herbs and roots that grow therein. The Toad loveth stinking places and dirty, and hateth places with good smell and odour, and so he fleëth out of the vineyard, when the vines begin to bloom, for he may not suffer nor sustain their good odour and smell. And these worms have double liver,—that one is most venomous, and that other is remedy, and is given instead of treacle against poison and venom; and for to assay and know which of these is good and which is evil, the liver is thrown into an ant-hill,—then the ants flee and [a]void the venomous part, and desire and choose that other part, and shall be taken and kept to the use of medicine. And in the right side of such a frog is a privy bone, that cooleth somedeal seething water, if it be thrown therein,—and the vessel may not heat afterward, but if the bone be first taken out; and witches use that bone to love and hate. And be that worm never so venomous, yet by burning he loseth the malice of venom, and taketh most virtue of medicine, and ashes thereof help wonderfully to recover flesh and skin that is haply lost, and to make sadness and sinews, and to healing and salvation of wounds, if the ashes be used in due manner.

Bartholomew (*Berthelet*), bk. xviii. § 17.

V. Frog.

A certain man, being in a garden with his Love, did take (as he was walking) a few leaves of sage, who rubbing his teeth and gums therewith, immediately fell down and died. Whereupon his said Love was examined how he died. She said, she knew nothing that he ailed, but that he rubbed his teeth with sage; and she went with the Judge and others into the garden and place where the same thing happened; and then she took of the same sage to shew them how he did, and likewise rubbed her teeth and gums therewith, and presently she died also, to the great marvel of all them that stood by. Whereupon the Judge, suspecting the cause of their deaths to be in the sage, caused the said bed of sage to be plucked and digged up, and to be burned, lest others might have the like harm thereby. And at the roots, or under the said sage, there was a great Toad found, which infected the same sage with his venomous breath. This may be a warning to such as use to eat raw and unwashed sage; therefore, it is good to plant rue round about sage, for Toads by no means will come nigh unto rue. *Lupton*, "Notable Things," bk i. § 1.

If you put a Toad in a new earthen pot, and the same be covered in the ground in the midst of a corn-field, there will be no hurtful tempests or storms there.

Ibid., bk. v. § 59.

The wise and learned men in old time did think, that a Toad put into a new earthen pot, and set it within the ground, and so covered with earth in the midst of a field, will drive away crows or birds from corn that is sowed there. But about harvest time they will that it be digged up, and so cast forth of the limits of the fields, lest the corn be bitter thereby. *Ibid.*, bk. vi. § 97.

There is a kind, like to the Toad of the water, but instead of bones it hath only gristles, and is bigger than the Toad of the fen, living in hot places. There is a little bone growing in their sides, that hath a virtue to drive away dogs from him that beareth it about him. All the winter-time they live under the earth, feeding upon earth, herbs and worms, and they eat earth by measure, for they eat so much every day as they can grip in their fore-foot, as it were sizing themselves, lest the whole earth should not serve them, till the spring. They also love to eat sage, and yet the root of sage is to them deadly poison. They destroy bees, without all danger to themselves, for they will creep to the holes of their hives, and there blow in upon the bees, by which breath they draw them out of the hive and so destroy them as they come out. About their generation, there are many worthy observations in nature; sometimes they are bred out of the putrefaction and corruption of the earth; it hath also been seen, that, out of the ashes of a Toad burnt, not only one, but many Toads, have been regenerated the year following. In the New World there is a province called Darien, the air whereof is wonderful unwholesome, because all the country standeth upon rotten marshes. It is there observed, that when the slaves, or servants water the pavements of the doors, from the drops of water which fall on the right hand are instantly many Toads engendered, as in other places such drops of water are turned into gnats. It hath also been seen that women conceiving with child have likewise conceived at the same time a frog, or a Toad, or a lizard. And for this cause, women, at such time as their child beginneth to quicken in their womb, do drink the juice of parsley and leeks, to kill such conceptions if any be. But in men's stomachs there are found frogs and Toads. This evil happeneth unto such men as drink water. And Toads are bred in the bodies of men, and yet afterwards these Toads do kill the bodies they are bred in; for the venom is so tempered, that at last it worketh when it is come to ripeness. For the casting out of such a Toad bred in the body, they take a serpent and [disem]bowel him; then they cut off the head and the tail; the residue of the body they likewise part into small pieces, which they seethe in water and take off the fat which swimmeth at the top, which the sick person drinketh, until by vomiting he avoid all the Toads in his stomach. Toads sometimes in anger lift up themselves, for great is their wrath, obstinacy, and desire to be revenged upon their adversaries, especially the red Toad; for if she take hold of any thing in her mouth, she will never let it go till she die, and many times she sendeth forth poison out of her buttocks or backer parts, wherewithal she infecteth the air, for revenge of them that do annoy her; and she knoweth the weakness of her teeth, and therefore she gathereth abundance of air into her body, wherewithal she greatly swelleth, and then, by sighing, uttereth that infected air as near the person that offendeth her as she can. A Toad useth one certain herb wherewithal it preserveth the sight, and also resisteth the poison of spiders, whereof I have heard this credible history related from the mouth of the good Earl of Bedford. It fortuned as the said Earl travelled in Bedfordshire, near unto a market-town called Owbourn [? Woburn], some of his company espied a Toad fighting with a spider, under a hedge; and the Earl saw how the spider still kept her standing, and the Toad divers times went back from the spider, and did eat a piece of an herb, which to his judgment was like a plantain; at the last, the Earl, having seen the Toad do it often, and still return to the combat against the spider, he commanded one of his men to go, and with his dagger to cut off that herb:—presently after the Toad returned to seek it, and, not finding it, swelled and broke in pieces. [This story is better told in *Lupton's* "Notable Things," bk. vi. § 30, but without the names, and with slight differences.] There was a monk in England who had in his chamber divers bundles of green rushes, wherewithal he used to strew his chamber at his pleasure; it happened on a day after dinner, that he fell asleep upon one of those bundles of rushes, with his face upward, and, while he there slept, a great Toad came and sat upon his lips, bestriding him in such manner as his whole mouth was covered. Now when his fellows saw it, they were at their wits' end, for to pull away the Toad was an unavoidable death, but to suffer her to stand still upon his mouth was a thing more cruel than death; and therefore one of them espying a spider's web in the window, wherein was a great spider, he did advise that the monk should be carried to that window, and laid with his face upward right underneath the spider's web. And as soon as the spider saw her adversary the Toad, she presently wove her thread, and descended down upon the Toad,—at the first meeting whereof the spider wounded the Toad, so that it swelled, and at the second meeting it swelled more, but at the third time the spider killed the Toad, and so became grateful to her host which did nourish her in his chamber; for at the third time the Toad leaped off from the man's mouth, and swelled to death. The mole also is an enemy to the Toad, for Albertus saw a Toad crying above the earth very bitterly, for a mole did hold her fast by the leg within the earth, labouring to pull her in again, while the other strove to get out of her teeth; and so on the other side, the Toads do eat the moles when they be dead. The cat doth kill serpents and Toads, but eateth them not, and unless she presently drink she dieth for it. The Toads of the earth are more poisonful than the Toads of the water, except those Toads of the water, which do receive infection or poison from the water, for some waters are venomous—but the Toads of the land, which do descend into the marshes, and so live in both elements, are most venomous; and the hotter the country is, the more full are they of poison. When an asp hath eaten a Toad, their biting is incurable, and bears, being killed by men after that they have eaten salamanders or Toads, do poison their eaters. A Toad hath two livers, and although both of them are corrupted, yet the one of them is full of poison, and the other resists poison. The spittle of Toads is venomous, for if it fall upon a man, it causeth all the hair to fall off from his head. Plantain and black hellebore, sea-crabs dried to powder and drunk, the stalks of dog's-tongue, the powder of the right horn of a hart, the milt, spleen and heart of a Toad, also the blood of the sea-tortoise mixed with wine, cummin and the rennet of a hare, also the blood of a tortoise of the land mixed with barley-meal, and the quintessence of treacle, and oil of scorpions,—all these things are very precious against the poison of serpents and Toads.

Topsell, "History of Serpents," pp. 726-30.

THE jaws of a Toad, sweating and foaming out poison, are not more dangerous than a pen.

Dekker's "Dead Term."

THOU shalt eat nothing but a poached spider, and drive it down with syrup of Toads.

"A Match at Midnight," i. 1.

Toad-stone.

The toad, ugly and venomous,
Wears yet a precious jewel in his head.
As You Like It, ii. 1, 13.

Some toads that breed in Italy and about Naples have in their heads a stone called a *crapo*, of bigness like a big peach, but flat, of colour grey, with a brown spot in the midst. *Batman*'s additions to *Bartholomew*, bk. xviii. § 17.

Noset, that is crapaudine, is a precious stone somedeal white, or of diverse colours. This stone is taken out of a toad's head, and is cleansed in the same head and in strong

wine and water, and sometime the shape of a toad seemeth therein with sharp feet and broad. This stone helpeth against biting of serpents and of creeping worms, and against venom,—for in presence of venom the stone warmeth and burneth his fingers that toucheth him.

Bartholomew (*Berthelet*), bk. xvi. § 17.

A Toad-stone, called *crapaudina*, touching any part bevenomed, hurt, or stung with rat, spider, wasps, or any other venomous beast, ceases the pain or swelling thereof.

Lupton, "Notable Things," bk. i. § 52.

A GOOD way to get the stone called *crapaudina* out of the toad: put a great, or overgrown toad (first bruised in divers places) into an earthen pot, and put the same in an ants' hillock, and cover the same with earth, which toad at length the ants will eat, so that the bones of the toad and stone will be left in the pot, which *Mizaldus* and many others hath oftentimes proved.

Lupton, "Notable Things," bk. vii. § 18.

YOU shall know whether the Toad-stone called *crapaudina* be the right and perfect stone or not: hold the stone before a toad, so that he may see it, and if it be a right and true stone, the toad will leap toward it, and make as though he would snatch it from you; he envieth so much that a man should have that stone. This was credibly told *Mizaldus* for truth by one of the French King's Physicians, which affirmed that he did see the trial thereof.

Ibid., bk. vii. § 79.

THERE is a precious stone in the head of a toad, and there be many that wear these stones in rings, being verily persuaded, that they keep them from all manner of gripings, and pains of the belly. But the art is in taking it out, for it must be taken out of the head alive, before the toad be dead, with a piece of cloth of the colour of red scarlet, wherewithal they are much delighted, so that while they stretch out themselves as it were in sport upon that cloth, they cast out the stone of their head, but instantly they sup it up again, unless it be taken from them through some secret hole in the said cloth, whereby it falleth into a cistern or vessel of water, into the which the toad dareth not enter, by reason of the coldness of the water. Now stones are engendered in living creatures two manner of ways, either through heat, or extreme cold, as in the snail, perch, crab, Indian tortoises and toads; so that by extremity of cold this stone should be gotten. In the presence of poison it will change the colour.

Topsell, "History of Serpents," p. 727.

[Topsell is neither for nor against the existence of this stone, but he cannot believe that it is generated by cold, because the stone is hard.]

Toadstool.

TROILUS AND CRESSIDA, ii. 1, 22.

OF MUSHROOMS OR TOAD-STOOLS.

SOME mushrooms grow forth of the earth; other upon the bodies of old trees. Many wantons that dwell near the sea, and have fish at will, are very desirous for change of diet to feed upon the birds of the mountains; and such as dwell upon the hills or champaign grounds do long after sea-fish; many that have plenty of both do hunger after the earthy excrescences called mushrooms. The mushrooms or Toadstools, which grow upon the trunks of old trees, very much resembling Jew's-ear, do in continuance of time grow unto the substance of wood, which the fowlers do call touchwood. This kind of mushroom is full of poison. With fuzz-balls, puck-fists and bull-fists in some places of England they use to kill or smoulder their bees, when they would drive the hives, and bereave the poor bees of their meat, houses and lives; these are also used in some places, where neighbours dwell far asunder, to carry and reserve fire from place to place. Poisonsome mushrooms groweth where old rusty iron lieth, or rotten clouts, or near to serpents' dens, or roots of trees that bring forth venomous fruit. Divers come up in April, others grow later about August. To conclude, few of them are good to be eaten, and most of them do suffocate and strangle the eater; therefore I give my advice unto those that love such strange and new-fangled meats to beware of licking honey among thorns, lest the sweetness of the one do not countervail the sharpness and pricking of the other. Fuzz-balls are noway eaten; the powder of them is fitly applied to merigalls, kibed heels and such like.

Gerard's "Herbal," *s.v.*

Tobacco.

[Though Shakespeare does not mention Tobacco, allusions to it are very frequent in the other dramatists of his time.]

OUR adulterate Nicotian or Tobacco, so called of the Knight Sir Nicot that first brought it over, which is the spirit's Incubus, that begets many ugly and deformed phantasies in the brain, which being also hot and dry in the second extenuates, and makes meagre the body extraordinarily. Of its own nature not sophisticate, it cannot be but a sovereign leaf for external maladious ulcers; and so it is for cacochymical bodies, and for the consumption of the lungs, and tisic, if it be mixed with colt's-foot dried. But as it is intoxicated and tainted with bad admixture, I must answer as our learned Paracelsus did, of whom my self did demand, whether a man might take it without impeachment to his health; who replied, as it is used, it must needs be very pernicious in regard of the immoderate and too ordinary whiff; for it will evacuate the stomach and purge the head for the present of many feculent and noisome humours, but after by his attractive virtue it proveth *Cæcias humorum*, leaving two ponds of water (as he termed them) behind it, which are converted into choler, one in the ventricle, another in the brain. And seeing every nasty and base *Tigellus* use the pipe, as infants their red corals, ever in their mouths, and many besides of more note and esteem take it more for wantonness than want, as Gerard speaks, I could wish that our generous spirits could pretermit the too usual, not omit the physical drinking of it. *Walkington*, "Optic Glass of Humours," pp. 105-7.

[There were different sorts of Tobacco. "Roll Trinidado, leaf and pudding," are mentioned in *Dekker's* "Gull's Hornbook."]

THIS is my friend Abel, an honest fellow,
He lets me have good Tobacco, and he does not
Sophisticate it with sack-lees or oil,
Nor washes it in muscadel and grains,
Nor buries it in gravel underground
Wrapp'd up in greasy leather, or piss'd clouts,
But keeps it in fine lily pots, that, opened,
Smell like conserve of roses, or French beans.
He has his maple block, his silver tongs,
Winchester pipes, and fire of juniper.

Ben Jonson, "The Alchemist," i. 3.

THREE pence a pipe-ful I will ha' made of all my whole half pound of Tobacco, and a quarter of a pound of colts-foot, mixed with it too, to eke it out.

Ibid., "Bartholomew Fair," ii. 2.

THE SECOND BILL:—If this city or suburbs of the same do afford any young gentleman of the first, second or third head, more or less, whose friends are but lately deceased, and whose lands are but new come into his hands, that (to be as exactly qualified as the best of our ordinary gallants are) is affected to entertain the most gentleman-like use of Tobacco: as first, to give it the most exquisite perfume: then, to know all the delicate sweet forms for the assumption of it: as also the rare corollary and practice of the Cuban ebullition, Euripus and whiff; which he shall receive, or take in here at London, and evaporate at Uxbridge, or farther, if it please him—If there be any such generous spirit, that is truly enamoured of these good faculties—may it please him but by a note of his hand to specify the place or Ordinary, where he uses to eat and lie,—and most sweet attendance with Tobacco and pipes of the best sort shall be ministered.

Ben Jonson, "Every Man out of his Humour," iii. 3.
Cf. "Return from Parnassus," iv. 1, and "Lingua," iv.

IN what Tobacco-shop in Fleet Street he takes a pipe of smoke in the afternoon.

Dekker, "Lanthorn and Candle-light."

YOUR pipe bears the true form of a woodcock's head.

"Every Man out of his Humour," iii. 9.

HERE'S a clean gentleman to receive.

"The Puritan," i. 4.

Scattergood: Please you to impart your smoke?
Longfield: Very willingly, sir.
Scattergood: In good faith, a pipe of excellent vapour.

[*Greene*, "Tu Quoque" (by John Cooke).

[See *Skelton's* "Ellinor Rumming."
Sixty thousand pounds' worth of tobacco was brought into England in 1610, according to Harcourt's "Description of Guiana."]

Topaz.

[Sir Topas, the curate's name in "Twelfth Night" (iv. 2) may be derived from this stone.]

TOPAZ was first found in an island of Arabia, in which island when the Troglodytes were diseased with hunger and tempest, they digged up roots of herbs, and they found this stone therewith. And if thou wipe this stone, thou darkest it, and if thou leavest him to his own kind, he is the more clear. And in treasury of kings nothing is more clear nor more precious than this precious stone, for clearness thereof taketh to himself the clearness of other precious stones that be about him, and he followeth the course of the moon, and as the moon is more full or less, so his effect is more or less, and helpeth against the passion lunatic [perhaps this is the reason of the curate's name in *Twelfth Night*, iv. 2], and stauncheth blood, and suageth fervent water, and suffereth it not to boil. It suageth both wrath and sorrow, and helpeth against evil thoughts and frenzy, and against sudden death.

Bartholomew (*Berthelet*), bk. xvi. § 96.

IF you wish that boiling water should overflow as soon as the hand is dipped into it, take the stone which is called Topaz; for it has been proved in our time that if the stone be placed in boiling water, it makes it flow away, so that if the hand be dipped into it, the stone may be drawn out at once—and this Parisius one of our brethren performed.

Albertus Magnus, "Of the Virtues of Stones."

IF the whites of many hens' eggs be taken, after a month they become glass, and hard as stone; and from this, first rubbed with saffron or red earth, the false Topaz is made.

Ibid., "Of the Wonders of the World."

Trout.

TWELFTH NIGHT, ii. 5, 25.

THIS fish of nature loveth flattery; for, being in the water, it will suffer itself to be rubbed and clawed, and so to be taken, whose example I would wish no maids to follow. *Cogan*, "Haven of Health" (1595), quoted by Steevens.

[Trout were baked and minced with dates, ginger, cinnamon, a quantity of sugar and butter, and served in three-cornered puffs (second part of "The Good Huswife's Jewel," p. 31).]

Turkey-cock.

i. KING HENRY IV. ii. 1, 28.

TURKEY-COCK, or cock of India, brought to us from India, or Arabia, or Africa. It seems to partake of the nature of the cock and of the peacock.

Minsheu's Dictionary, *s.v.*

A TURKEY or Guinea hen. *Ibid.*

FIRST brought to England in Henry VIII.'s reign.

Malone in *Steevens'* Shakespeare.

HENS, and especially those of Turkey or the Indies seem plentifully served in the [Italian] markets.

Fynes Moryson, "Itinerary," part iii. p. 110.

THE state of a fat Turkey, the decorum
He marches in with, all the train and circumstance;
'Tis such a matter, such a glorious matter,—
And then his sauce with oranges and onions,
And he display'd in all parts! For such a dish now
And at my need I would betray my father.

Beaumont and Fletcher, "Woman Pleased," iii. 1.

TURKEY-PIE.

Ben Jonson, "Bartholomew Fair," i. 6, and *cf.* "Return from Parnassus," ii. 6.

Turnip.

MERRY WIVES OF WINDSOR, iii. 4, 90.

THE Turnip groweth in fields and divers vineyards or hop-gardens in most places of England. The small Turnip groweth by Hackney, and those that are brought to Cheapside Market from that village are the best that ever I tasted. Turnips flower and seed the second year after they are sown; for those which flower the same year that they are sown are a degenerate kind, called in Cheshire about the Namptwitch [*i.e.*, Nantwich] Mad Neeps—of their evil quality in causing frenzy and giddiness of the brain for a season. The root is many times eaten raw, especially of the poor people in Wales, but most commonly boiled or roasted or baked; the young and tender shoots or springs of turnips at their first coming forth of the ground boiled and eaten as a salad.

Gerard's "Herbal," *s.v.*

WITH us the yellow, which comes from Denmark, is preferred; by others, the red Bohemian. The stalks of the common Turnip, when first beginning to bud, being boiled, eat like asparagus.

Evelyn, "Acetaria," § 79.

THE best husbandmen would have the seedsmen [of Turnips or rapes] to be naked when he sows them, and in sowing to protest, that this which he doth is for himself and his neighbours. Mark how many days old the moon was when the first snow fell the winter next before,—for if a man do sow rapes or Turnips within the foresaid compass of that time, the moon being so many days old, they will come to be wondrous great, and increase exceedingly.

Holland's Pliny, bk. xviii. ch. xiii.

THE Turnip or navew groweth sometimes to thirty or forty pound weight. The best way of use is accounted, first to boil them, and, the water being poured out, then to boil them again with fat beef, adding to them some pepper.

Hart, "Diet of the Diseased," bk. i. ch. xiii.

Turquoise.

MERCHANT OF VENICE, iii. 1, 126.

TURQUOISE is a white yellow stone, and hath that name of the country of Turkey, there it is bred. This stone keepeth and saveth the sight, and breedeth gladness and comfort.

Bartholomew (*Berthelet*), bk. xvi. § 97.

THE Turquoise is formed beyond the farthest parts of India among the inhabitants of the mountain Caucasus, [and in] Carmania. They be found in icy cliffs hardly accessible, where you shall see them bearing out after the manner of bosses like unto eyes. *Holland's Pliny*, bk. xxxvii. ch. viii.

A TRUE wife should be like a Turquoise stone, clear in heart in her husband's health, and cloudy in his sickness.

Alex Nicholas, "Discourse of Marriage and Wiving," ch. xiv. § 18.

AND true as Turquoise in the dear lord's ring
Look well or ill with him.

Ben Jonson, "Sejanus," i. 1.

THE Turquoise, which who haps to wear
Is often kept from peril.

Drayton, "Muses' Elysium."

THE Turquoise doth move, when there is any peril prepared to him that weareth it.

Edw. Fenton, "Secret Wonders of Nature."

THE Turquoise is likewise said to take away all enmity, and to reconcile man and wife. *Thos. Nicols*, "Lapidary."

[These three quotations are from Steevens' notes to the passage in "Merchant of Venice."]

Turtle, Turtle-dove.

WINTER'S TALE, iv. 4, 154.
MERRY WIVES OF WINDSOR, iii. 3, 44.
i. KING HENRY VI., ii. 2, 30.

THE Turtle hath that name of the voice, and is a simple bird as the culvour, but is chaste, far unlike the culvour, and if he loseth his make [*i.e.*, mate], he seeketh not company of any other, but goeth alone, and hath mind of the fellowship that is lost, and groaneth alway, and loveth and chooseth solitary places, and flieth much company of men

[*Troilus and Cressida*, iii. 2, 185, and *Winter's Tale*, *ut supra*]. He cometh in springing time and warneth of novelty of time with groaning voice. And in winter he loseth his feathers, and then he hideth him in hollow stocks. And against summer, in springing time, when his feathers spring again, he cometh out of his hole in the which he was hid, and seeketh convenable place and stead for to breed in. The Turtle layeth eggs twice in springing time, and not the third time, but if the first eggs be corrupt. Also the blood of her right wing is medicinable, as the blood of a swallow, and of a culvour or dove.

Bartholomew (*Berthelet*), bk. xii. § 34.

AMATIDES is a precious stone; if a cloth be touched therewith, the cloth withstandeth fire and burneth not, though it be put therein; but it receiveth brightness and seemeth the more clear. And withstandeth all evil doing of witches. *Ibid.*, bk. xvi. § 19.

IT is supposed that in the maw of the cock Turtle-dove this stone is to be found, and hath virtue to increase concord and love.

Batman's addition to *Bartholomew*, *ut supra.*

IF the heart of a Turtle-dove be worn in the skin of a wolf, the wearer will never thenceforth be wanton. If its heart be burnt, and put on the eggs of any bird, never will it be possible that they should be hatched. And if its feet be hung on a tree, from thenceforth it will not bear fruit. And if a hairy place be anointed with its blood, and the water in which a mole has been boiled, the black hairs will fall off. *Albertus Magnus*, "Of the Virtues of Animals."

Unicorn.

A living drollery! Now I will believe
That there are unicorns.

TEMPEST, iii. 2, 22.

JULIUS CÆSAR, ii. 1, 204.

AN Unicorn is a right cruel beast, and hath that name for he hath in the middle of the forehead an horn of four foot long; and that horn is so sharp and so strong, that he throweth down all, or thirleth [pierceth] all that he reseth [rageth] on. And this beast fighteth oft with the elephant. And the Unicorn is so strong, that he is not taken with might of hunters; but a maid is set there as he shall come, and she openeth her lap, and the Unicorn layeth thereon his head, and leaveth all his fierceness, and sleepeth in that wise, and is taken as a beast without weapon, and slain with darts of hunters. The Unicorn froteth [rubs] and fileth his horn against stones, and sharpeth it, and maketh it

ready to fight in that wise. And his colour is bay. There be many kinds of Unicorns, for some be Rhinoceros [*q.v.*], and some *Monoceron*, and *Ægloceron*. And *Monoceron* is a wild beast, shaped like to the horse in body, and to the hart in head, and in the feet to the elephant, and in the tail to the boar, and hath heavy lowing, and an horn strutting in the middle of the forehead of two cubits long. And in Ind be some one-horned asses, and such an ass is called *Monoceros*, and is less bold and fierce than other Unicorns. And *Ægloceron* is a manner of Unicorn, that is

a little beast like to a kid. Also in Ind be one-horned oxen with white specks and bones, and with thick hoofs as horses have.
Bartholomew (*Berthelet*), bk. xviii. § 90.

TRADES that lay dead and rotten, and were in all men's opinion utterly damned, started out of their trance, as though they had drunk of Aqua Cœlestis, or Unicorn's horn, and swore to fall to their old occupations.
Dekker, "The Wonderful Year 1603."

SOME hunt the Unicorn for the treasure on his head, and they are like covetous men, that care not whom they kill for riches.
Ibid., "Lanthorn and Candlelight," ch. iii.

IN St. Mark's church they will show you two Unicorns' horns, of which the red is the male, and the yellow the female.
"A true Description of what is most worthy to be seen in Italy," etc. (*circa* 1590).

[Topsell ("Four-footed Beasts," pp. 551, 552) takes those who do not believe in the Unicorn very seriously to task for their unbelief, not to say atheism. He inclines to the belief that the Unicorn is the wild ass of India, but is not sure, because "the feet of the wild asses are whole and not cloven like the Unicorn's, and their colour white in their body, and purple on their head; and the horn differeth in colour from the Unicorn's, for the middle of it is only black, the root of it white, and the top of it purple; and the Indians of that horn do make pots, affirming that whosoever drinketh in one of those pots shall never take disease that day, and, if they be wounded, shall feel no pain, or safely pass through the fire without burning, nor yet be poisoned in their drink, and therefore such cups are only in the possession of their kings, neither is it lawful for any man except the king to hunt that beast. Now in the kingdom of Basman, which is subject to the great Cham, there are Unicorns somewhat lesser than Elephants, having hair like oxen, heads like boars, feet like elephants, one horn in the middle of their foreheads, and a sharp, thorny tongue, wherewith they destroy both man and beast, and they muddle in the dirt like swine. In a certain region of the new-found world, under the equinoctial, there is a living creature, with one horn (which is crooked and not great), having the head of a dragon, and a beard upon his chin, his neck long and stretched out like a serpent's, the residue of his body like to a hart's, saving that his feet, colour, and mouth, are like a lion's."]

THE horn, growing out of the forehead betwixt the eyelids is neither light nor hollow, nor yet smooth like other horns, but hard as iron, rough as any file, revolved into many plights, sharper than any dart, straight and not crooked, and everywhere black except at the point. His horn, being put into the water, driveth away the poison, that he may drink without harm, if any venomous beast shall drink therein before him. This cannot be taken from the beast being alive, forasmuch as he cannot possibly be taken by any deceit. The horn of this beast being put upon the table of kings, and set among their junkets and banquets, doth bewray the venom (if there be any such therein) by a certain sweat which cometh over it. There are found in Europe to the number of twenty of these horns pure, and so many broken.

These beasts are very swift, and their legs have no articles. There was nothing more horrible than the voice or braying of it, for the voice is strained above measure. He feareth not iron, nor any iron instrument. He is an enemy to the lions, wherefore as soon as ever a lion seëth an Unicorn, he runneth to a tree for succour, that so when the Unicorn maketh force at him, he may not only avoid his horn, but also destroy him; for the Unicorn in the swiftness of his course runneth against the tree, wherein his sharp horn sticketh fast; then when the lion seëth the Unicorn fastened by the horn, without all danger he falleth upon him and killeth him. These things are reported by the King of Ethiopia in an Hebrew epistle unto the Bishop of Rome. [Topsell describes the way of catching the Unicorn given in the quotation above from Bartholomew, and adds: "Concerning this opinion we have no elder authority than Tzetzes, who did not live above five hundred years ago, and therefore I leave the Reader to the freedom of his own judgement to believe or refuse this relation."] Rich men do usually cast little pieces of this horn in their drinking-cups, either for the prevention or curing of some certain disease. It being cast in wine doth boil. [He avers that spurious Unicorn's horn, made of ivory, was sold by apothecaries and others.] The price of that which is true is reported at this day to be of no less value than gold. For experience of the Unicorn's horn, to know whether it be right or not,—put silk upon a burning coal, and upon silk the aforesaid horn, and if so be that it be true the silk will not be a whit consumed.

Topsell, "Four-footed Beasts," pp. 551-9.

WE are so far from denying that there is any Unicorn at all, that we will affirm there are many kinds thereof. In the number of quadrupeds, we will concede no less than five; that is, the Indian ox, the Indian ass, the rhinoceros, the oryx, and that which is more eminently termed *Monoceros* or *Unicornis*. Some in the list of fishes; and some Unicorns we will allow even among insects [here follow two folio pages of argument about the origin and genuineness of the horns].

Sir Thos. Browne, "Vulgar Errors," bk. iii. ch. xxiii.

[*Fynes Moryson*, in his "Itinerary," describes "two whole Unicorns' horns, each more than four foot long, and a third, shorter," which were in the Treasury of St Mark at Venice (part i. bk. ii. ch. i.), and "great Unicorns' horns, and the chief kinds of precious stones" in Naples (*ibid.*, ch. ii.).]

THE Unicorn is hunted for his horn,
The rest is left for carrion.

Middleton and Rowley, "A Fair Quarrel," iii. 2.

OF the Unicorn none hath been seen these hundred years last past. *Purchas*' "Pilgrims," p. 502 (ed. 1616).

[But the ingenious gentlemen who edited the "British Apollo" would not go so far as to deny (in 1710) the existence of the Unicorn.]

Urchin.

TEMPEST, i. 2, 326.

V. Hedgehog.

Venom.

SOME [beasts] have slaying tongues and venomous, through malice and wodeness of the humour that hath mastery therein; as the tongues of serpents, adders, dragons, and of a wode hound, whose biting is most venomous, his tongue droppeth Venom, and corrupteth and infecteth the water in which it falleth in, and who that drinketh of that water shall become wode. And the tongues of adders be black, blue or reddish, speckled, sharp, and in moving most swift, and that happeneth through the wode and venomous humour, the which so swiftly moveth the tongue that one tongue seemeth forked and twisted. And though the tongue of an adder [asp] is full of deadly Venom, while it liveth in the body of the adder, yet, when it is taken from the body of the adder, and dried, it loseth the Venom, and by it is known when Venom is present,—therefore in the presence of Venom, such a tongue useth to sweat,—therefore such a tongue is accounted precious among treasures of kings. *Bartholomew* (*Berthelet*), bk. v. § 21.

Over and above the foresaid evils and passions, most perilous death and evils hap and come to mankind by wicked Venom. And for all kind of Venom is contrary to the complexion of mankind, it slayeth suddenly, but [unless] men have the sooner help and remedy. Some Venom cometh of corruption of meat and drink; and some of biting of creeping worms and of adders, and of serpents, and of other beasts, of whom their humours and teeth be venomous to man's body. Also some Venom is hot and dry, as the Venom of an adder that hight viper, and other such; and some Venom is cold and dry, as the Venom of scorpions; and some Venom is cold and moist, as the Venom of attercops [spiders]. And the Venom of males is more sharp and strong than the Venom of females, and yet the female serpents have more teeth than males, and therefore they be taken for the worse. Also the Venom of the old serpents is worse than the Venom of the young; and of great and long worse than of the short of the same kind. Also the Venom of them that abide in hills and woods is worse than of them which be nigh cliffs and banks of waters. Venom of a cockatrice is so violent that it burneth all thing, which it nigheth; and so about his den and his hole nothing waxeth green. One touched such a worm with his spear in India, and forthwith fell down dead, and his horse also. Also the Venom of a dragon is full malicious, and his Venom is most in the tail, and in the gall. *Bartholomew* (*Berthelet*), bk. vii. § 66.

WHOSOEVER is stricken or hurt of any venomous worm, or other thing, or else bitten with a mad dog, let them take heed diligently that the same thing that did hurt them see them not until they be perfectly whole. For the Hebrew Physicians say that the party hurt shall then die, or else be in peril afresh; yea, though they begin to wax whole when they see them.

Lupton, "Notable Things," bk. v. § 72.

Vine.

COMEDY OF ERRORS, ii. 2, 176.

VINES be perched and railed and bound to trees that be nigh to them. The crooks of the Vines holdeth things that be nigh thereto, for [so that] boughs and branches of the Vine should not be slacked far for the succour, and shaken, and disparpled [*or* "disparkled," *i.e.*, "scattered"], and hurled with blasts of wind, but they should so come to bear and save the fruit without peril. Rain gendereth and breedeth certain worms and malshrags [caterpillars] and snails, that grow and fret burgeoning and leaves of the Vine, and leaveth lightly the Vine so spoiled, gnawn and eaten; and this evil breedeth in moist time, easy and soft. And of evil blasts of winds cometh and breedeth as it were cob-webs, and beclippeth [surrounds] and wasteth the fruit, and burneth and grieveth it. Also the Vine hateth the radish, and all manner cole, and hateth also hazels, for when such be nigh to the Vines, then the Vines be ailing and sick, and nitre—much like to salt—alum and sea-water, and beans, and vetches, and namely [especially] in the last, cutting be venom to Vines, and destroy them. [*Bartholomew—fabæ ac viciæ putamina ultima et maxime interimentia vitium sunt venena.*] And in some parts and countries be so great Vines, that they make images, posts and stocks of Vines; as it fareth in the image and mammet [idol—from Mahomet] Jupiter in the city of Populonia. And men stied [climbed —*Bartholomew*] upon a Vine to the top of the temple of Diana of Ephesus. Also posts and pillars made of such Vines dure and last without corruption long time. The juice [of the Vine] with oil laid to an hairy place in a plaster-wise doth away the hair. The rind of the Vine doth away warts. Moreover the ashes of the Vine healeth with oil stinging of scorpions, and biting of hounds. Ashes of the rind by itself restoreth and multiplieth hair that is fallen.

Bartholomew (*Berthelet*), bk. xvii. § 177.

[Evelyn ("Kalendarium Hortense") mentions the following sorts: Amboise, Frontignac (Grizzling, excellent, White, excellent, Blue), Burgundian, Early Blue, Muscatel (Black, White, excellent), Morillon, Chassela, Cluster-grape, Parsley, Raisin, Bursarobe, Burlet, Corinth, Large Verjuice (excellent for sauces and salleting).]

SOMETIMES there hath been tendrils of gold found in the Vine; whereof there hath been money coined. And in Germany, within Danubia, Vines did bear little nails and leaves of pure gold, which were given as presents to kings and dukes.

Lupton, "Notable Things," bk. iv. § 42.

I MUSE not a little wherefore the planting of Vines should be neglected in England.

Holinshed, "Description of Britain," p. 110.

WILLIAM OF MALMESBURY writes that Gloucestershire yielded in his time plenty of Vines, abounding with grapes of a pleasant taste, so as the wines made thereof were not sharp, but almost as pleasant as the French wines; which Camden thinks probable, there being many places still called Vineyards, and attributes it rather to the inhabitants' slothfulness, than to the fault of the air or soil, that it yields not wine at this day.

Fynes Moryson, "Itinerary," part iii. pp. 138-9.

V. Grape, Raisin, Wine.

Vinegar.

ii. KING HENRY IV., ii. 1, 102.

WINE is first sweet and temperate in savour, and is corrupt by long working of the sun or of the air, and by long boiling, and turneth into sourness when it hath no virtue by the which it may be kept and saved. And by subtlety of the substance thereof, and by feebleness of the coldness it thirleth [traverses] the body soon, and cometh to the well worse place. And Vinegar helpeth against venom and also against venomous beasts which slayeth.

And strong Vinegar done upon iron or upon the cold ground boileth and seetheth anon. Also Vinegar stauncheth parbreaking [vomiting] and wambling, if the mouth and the other part of the throat be washed therewith, and thrown out again; and helpeth deaf ears, and openeth the hearing and the ways; and sharpeth the sight of eyes. And drasts [dregs] of Vinegar helpeth against the biting of a wode hound, and of the crocodile.

Bartholomew (*Berthelet*), bk. xviii. § 188.

"How to make white of red Vinegar:—Fetch your Vinegar at St. Katherine, a groat a gallon"—[add a pottle of elder flowers to six gallons of Vinegar. Renew every year with fresh flowers and Vinegar ("Good Huswife's Treasury," bk. vi.)].

EVEN now I strike his body to wound:
Behold, now his blood springs out on the ground.
(Stage-direction: *A little bladder of Vinegar pricked.*)

"Lamentable Tragedy of Cambyses";
so "Return from Parnassus," i. 2.

I WILL sell her
For twopence a quart, Vinegar! Vinegar in a wheelbarrow!

Randolph, "Hey for Honesty," etc., iv. 3.

Is your patent for making Vinegar confirm'd?

Chapman, "The Ball," ii. 2.

Vineyard.

MEASURE FOR MEASURE, iv. 1, 29, etc.

V. Vine.

A VINEYARD is busily tilled and kept, and oft visited and overseen of the earth-tillers, and keepers of vines, that they be not appaired [damaged] neither destroyed with beasts, and a wait is there set in an high place to keep the Vineyard, that the fruit be not destroyed, and is left in winter without keeper or waiter. The smell of the Vineyard that bloometh is contrary to all venomous things, and therefore adders and serpents flee, and toads also, and may not sustain and suffer the noble savour thereof. Foxes lurk and hide themself under vine-leaves, and gnaw covetously and fret the grapes of the Vineyard, and namely when the keepers and wardens be negligent and reckless; and it profiteth not that some unwise men doth, that close within the Vineyard hounds, that be adversaries to foxes, for few hounds so closed waste and destroy more grapes than many foxes should destroy, that come and eat thievishly. Therefore wise wardens of Vineyards be full busy to keep that no swine nor tame hounds nor foxes come into the Vineyard.

Bartholomew (Berthelet), bk. xvii. § 180.

[Holinshed ("Description of Britain," p. 111) notes that there were enclosed parcels [*i.e.*, of land] almost in every abbey yet called the vineyards, that Smithfield in the reign of King Stephen was a profitable vineyard, and that the Isle of Ely was in the first times of the Normans called L'Isle des Vignes.]

Violet.

VIOLET is a little herb in substance, and the flower thereof smelleth most, and so the smell thereof abateth the heat of the brain, and refresheth and comforteth the spirits of feeling, and maketh sleep. And the more virtuous the flower thereof is, the more it bendeth the head thereof downward. *Bartholomew (Berthelet)*, bk. xvii. § 191.

VERY many by these Violets receive ornament and comely grace; for there be made of them garlands for the head, nosegays and posies; yea gardens themselves receive by these the greatest ornament of all, chiefest beauty and most gallant grace; for they admonish and stir up a man to that which is comely and honest. The seed is good against the stinging of scorpions.

Gerard's "Herbal," *s.v.*, where he describes the purple garden Violet, the white, the double purple (or white), the yellow (wild or mountain), and the dog Violet.

I THINK of kings' favours as of a marigold flower, or like the Violets in America, that in summer yield an odoriferous smell, and in winter a most infectious savour.

"A Knack to Know a Knave."

[Onions and Violets were conserved or pickled together in alternate layers (Dawson, "Good Huswife's Jewel," 1596).]

VIOLET is for faithfulness
  Which in me shall abide;
Hoping likewise that from your heart
  You will not let it slide.

"A Handful of Pleasant Delights" (1584), quoted by Malone.

Viper.

I am no viper, yet I feed
On mother's flesh which did me breed.

PERICLES, i. 1, 64.

VIPER is a manner kind of serpents that is full venomous, and hath that name for he bringeth forth brood by strength; for when her womb draweth to the time of whelping, the whelps abideth not convenable time nor kind passing, but gnaweth and fretteth the sides of their mother, and they come so into this world with strength and with the death of the mother. The male doth his mouth into the mouth of the female, and she waxeth wode in liking, and biteth off the head of the male, and so both father and mother are slain. Of this serpent be made pasties [lozenges], of the which is made treacle, that is remedy against venom. And this adder Viper sustaineth and may bear hunger long time in a strong winter; and cometh to the den under earth, and casteth first away his venom, and doth sleep there until springing-time come again. And when the pores of the earth open, then by heat of the sun this serpent Viper awaketh and cometh out of his den, and for his sight is appaired [impaired] by the long abiding under the earth, he seeketh the root of fennel or the herb of it, and washeth his dim eyes with the juice thereof to recover his sight which he hath lost. And *Tyrus* is a manner serpent that hight Viper also, and great serpents flee this serpent *Tyrus*, though he be little; and all his body is rough, and when he biteth any thing, all that is about the thing rotteth anon. And among all serpents the kind of Viper is worst, and when he would gender, he wooeth a manner lamprey, and cometh to the brink of the water, that he troweth that lamprey is in, and calleth her to him with hissing, and exciteth and wooeth her to byclipping [embracing]; and this lamprey cometh anon; and anon as the Viper seeth that she is ready, he casteth away all his venom, and goeth then, and byclippeth the lamprey; and when the deed is done, then he drinketh and taketh again the venom which he had cast away, and so turneth again to his den with his venom. Also this adder Viper swalloweth a certain stone, and some men [Scythians] knoweth that, and openeth slily the serpent, and taketh out that stone, and useth it against venom. Also if the dragon or the adder which hight asp biteth a man or a beast, the head of the adder Viper healeth him and saveth him, if it be laid to the wound. And againward, the flesh of the adder asp ofttimes healeth and saveth him that the adder viper stingeth, and draweth out the venom. *Bartholomew* (*Berthelet*), bk. xviii. § 117.

The Viper hath no ears. It conceiveth at the mouth. Vipers sometimes eat scorpions, and in Arabia they not only delight in the sweet juice of Balsam, but also in the shadow of the same; but above all kinds of drink they are most insatiable of wine. It is certain and well known what great enmity is betwixt mankind and Vipers, for the one always hateth and feareth the other; wherefore, if a man take a Viper by the neck, and spit in his mouth, if the spittle slide down into his belly, it dieth thereof and rotteth as it were in a consumption. Vipers also are enemies to oxen, also to hens and geese, likewise to the dormouse; when the Viper cometh to the nest of a dormouse, and findeth there her young ones, she putteth out all their eyes, and afterwards feedeth them very fat, yet killeth every day one, as occasion of hunger serveth; but if in the meantime a man or any other creature do chance to eat of those dormice, whose eyes are so put out by the Viper, they are poisoned thereby. There is a kind of harmless serpent called *Parea*, which is an enemy unto Vipers and killeth them. A Viper climbed up into a tree to the nest of a magpie, whereupon the old one was sitting; this poor pie did fight with the Viper, until the Viper took her fast by the thigh, so as she could fight no more, yet she ceased not to chatter and cry out to her fellows to come and help her, whereupon the male pie came, and seeing his female so gripped by the Viper, he ceased not to peck upon his head until the brains came out, and so the Viper fell down dead. The scorpions and the Vipers are enemies one to another. The tortoise of the earth is also an enemy to the Viper, and the Viper to it, wherefore if it can get origan, or wild savory, or rue, it eateth thereof, and then is nothing afraid to fight with the Viper, but if the tortoise can find none of these, then they die incontinently by the poison of the Viper, and of this there hath been trial. Garlic is poison to the Viper, and therefore having tasted thereof she dieth, except she eat some rue. A Viper being struck with a reed once, it amazeth her, and maketh her senseless, but being struck the second time, she recovereth and runneth away; and the like is reported of the beech-tree, saving that it slayeth the Viper, and she is not able to go from it. If you lay fire on the one side, and a piece of yew on the other side, and then place a Viper in the middle betwixt them both, she will rather choose to run through the fire, than to go over the branches of yew. The Viper is also afraid of mustard-seed, for, it being laid in her path, she flieth from it, and, if she taste of it, she dieth. If the hands or the body of a man be anointed with the juice of the root of Arum, the Viper will never bite him; the like is reported of the juice of Dragons, expressed out of the leaves, fruit, or root. Also if a Viper do behold a good smaragd [emerald], her eyes will melt and fall out of her head. But the Viper is most delighted with vetches and the savine-tree. When the male misseth the female, he seeketh her out very diligently, and with a pleasing and flattering noise calleth for her, and when he perceiveth she approacheth, he casteth up all his venom, as it were in reverence of matrimonial dignity. In Egypt they eat Vipers and divers other serpents, with no more difficulty than they would do eels, so do many people both in the Eastern and Western parts of the New-found-lands. Whose diet of eating Vipers I do much pity, if the want of other food constrain them thereunto; but if it arise from the insatiable and greedy intemperancy of their own appetites, I judge them eager of dainties, which adventure for it at such a market of poison. A mountain-viper chased a man so hardly that he was forced to take a tree, unto the which when the Viper was come, and could not climb up to utter her malice upon the man, she emptied the same upon the tree, and by and by after, the man in the tree died, by the savour and secret operation of the same. If a man chance to tread upon the reins of a Viper unawares, it paineth him more than any venom, for it spreadeth itself over all the body incurably; and the mushrooms or toad-stools which grow near the dens and lodgings of Vipers are also found to be venomous. The eating of Vipers is an admirable remedy against leprosy. [He narrates several instances, in one of which a leprous woman was not only cured of her leprosy, but "soon after conceived a man-child, having been barren before the space of 40 years."] The skin of the Viper beaten to powder and laid upon the places where the hair is fallen, it doth wonderfully restore hair again. *Topsell*, "History of Serpents," pp. 800-10.

It would infect the genius of the air
With mists contagious (as if compos'd
Of Viper-steam).

Glapthorne, "The Hollander," iii. 1.

V. Serpent.

Vulture.

Merry Wives of Windsor, i. 3, 94.

The Vulture hath that name of slow flight, for of the plenteousness of much flesh, he lacketh swiftness of flight. And some men tell that she conceiveth and is conceived, and gendereth and is gendered without joining of treading; and they tell that they live an hundred years. This bird is cruel about his own birds, as the kite is; and if she seeth her birds too fat, she beateth them with her feet and bill to make them lean. Also in this bird the wit of smelling is best; and therefore by smelling he savoureth carrions that be far from him, that is beyond the sea; and againward [*i.e.*, *vice versâ*]. And the Vulture fighteth with the gentle falcon, and fleëth about him, and when he hath overcome him he dieth. And he hunteth from mid-day to night; and resteth still from the sun-rising to that time. And when he ageth, his over bill waxeth long and crooked over the nether, and dieth at the last for hunger. Also the Vulture is a much stinking fowl, and unclean. And the Vulture is contrary to serpents;—for if his feathers be burnt, the smell thereof driveth away serpents. And the heart thereof maketh a man sicher [secure] and safe that beareth it among serpents and wild beasts. The heart bound in a lion's skin, or in a wolf's skin, driveth away fiends. His right foot bound to the left foot healeth that acheth; the left foot also healeth the right foot. His tongue plucked out with iron, and hanged about a man's neck in new cloth, maketh a man gracious to get of a man what he desireth. *Bartholomew* (*Berthelet*), bk. xii. § 35.

Wall-newt.

KING LEAR, iii. 4, 135.

V. Lizard.

Walnut.

MERRY WIVES OF WINDSOR, iv. 2, 170.

IN great French nuts generally the shape of the cross is printed within, as they know well that take heed thereto.

Bartholomew (*Berthelet*), bk. xvii. § 108.

DRY Nuts taken fasting with a fig and a little rue withstand poison, prevent and preserve the body from the infection of the plague. The green and tender nuts boiled in sugar, and eaten as sucket, are a most pleasant and delectable meat, and expel poison. The oil of Walnuts made in such manner as oil of almonds maketh smooth the hands and face, and taketh away scales or scurf, black and blue marks that come of stripes or bruises. With onions, salt and honey, they are good against the biting of a mad dog or man, if they be laid upon the wound.

Gerard's "Herbal," *s.v.*

[A recipe for confection of Walnuts is given in the second part of "The Good Huswife's Jewel," p. 40.]

ADVERSE and contrary Walnuts are to the nature of onions, and do keep down and repress their strong smell which riseth from them, after a man hath eaten them. The shell of a Walnut is good to burn or sear an hollow tooth; the same being burnt, pulverised and incorporate with oil or wine serveth to anoint the heads of young babes for to make the hair grow thick; and in that manner it is used to bring the hair again of elder folk, when through some infirmity it is shed.

Holland's Pliny, bk. xxiii. ch. xviii.

IF an oak be set near unto a Walnut-tree, it will not live. *Ibid.*, bk. xxiv. ch. I.

Warden.

WINTER'S TALE, iv. 3, 20.

FRUITS in Prime or yet lasting:—*Pears*:—Warden (to bake); white, red and French Wardens (to bake or roast), etc.

Evelyn's "Kalendarium Hortense," November and December.

LEWES Warden (best without compare).

Ibid., "Catalogue of Pears."

WARDEN in mince-pies.

"Good Huswife's Treasury."

To conserve Wardens all the year in syrup.

Second part of "The Good Huswife's Jewel," p. 38.

FOR tart-stuff either Wardens, barberries, or damsons.

"Good Huswife's Treasury," p. 7, a.

I WOULD have him roasted like a Warden
In brown paper.

Beaumont and Fletcher, "Cupid's Revenge," iv. 2.

Wasp.

TAMING OF THE SHREW, ii. I, 210, etc.

HE is a political and flocking or gregal creature, subject to monarchy, of a very quarrelsome disposition, and very prone to choler. Isidore affirms that Wasps come out of the putrefied carcase of asses, although he may be mistaken; for all agree that the scarabees [beetles] are procreated from them; rather I am of opinion that they are sprung from the dead bodies of horses,—for the horse is a valiant and warlike creature. Other sorts of them are produced out of the putrid corpse of the crocodiles. Wasps come out of the putrefaction of an old deer's head, flying sometimes out of the head, sometimes out of the nostrils. Also Wasps are begotten of the earth and rottenness of some kind of fruits. The Wasps called Ichneumons are less than the rest; they kill spiders, and carry them into their nests, and daub them over with dirt, and so sitting upon them do procreate their own species. Of the Wasps as well wild as tame some have no sting; also very many of them that have stings lose them upon the approach of winter. They feed on flesh of serpents and then they sting mortally. They themselves are a plaister for their own stings. The distilled water of common Wasps applied to the belly makes it swell as if it had the dropsy, by which trick [the guileless man is deceived by the designing woman]. The Wasp will not come near any man that is anointed with oil and the juice of mallows.

Thos. Mouffet, "Theatre of Insects," pp. 921-6.

THE Wasp scorns that flower from which she hath fetched her wax. "Euphues' Golden Legacy."

WASPS feeding on serpents make their stings more venomous. *Lilly*, "Sappho and Phaon," iii. 3.

V. Fly.

Water-rat.

MERCHANT OF VENICE, i. 3, 23.

WATER-RAT. *V.* Otter. *Minsheu's* Dictionary, *s.v.*

THE Water-rat hunteth fishes in the winter.

Topsell, "Four-footed Beasts," p. 404.

Water-rug.

MACBETH, iii. 1, 94.

[Probably a rough-haired water-dog; perhaps a water-spaniel.]

Water-spaniel.

TWO GENTLEMEN OF VERONA, iii. 1, 271.

THEY use to shear their hinder-parts, that so they may be the less annoyed in swimming.

Topsell, "Four-footed Beasts," p. 122.

WITH these dogs we fetch out of the water such fowl as be stung to death by any venomous worm.

Ibid., p. 134.

Wax.

WAX is the drasts [dregs] of honey, and within the substance thereof be gathered the liquors of honey meddled with the drasts of Wax. Tables be filled and dressed with Wax, simple or coloured, and therein be letters and divers figures and shapes written or planed by the office of pointels [points]. And for divers uses linen cloths be waxed. And Wax keepeth and saveth books from rain and from water. *Bartholomew* (*Berthelet*), bk. xix. § 61.

He's a man of wax.
ROMEO AND JULIET, i. 3, 75.

I CANNOT blame you for loving of Sophos;
Why, he's a man as one should picture him in Wax.

"Wily Beguiled."

O FOOT! O leg! O hand! O body! face!
By Jove, it is a little man of Wax!

Nat. Field, "A Woman is a Weathercock," i. 2.

WHEN he is in his scarlet clothes, he looks like a man of Wax, and I had as lief have a dog o' Wax; I do not think but he lies in a case o' nights.

Ibid., "Amends for Ladies," iii. 3.

Weasel.

AS YOU LIKE IT, ii. 5, 13.

THE Weasel is as it were a long mouse. This beast hath a guileful wit, and nourisheth her kittens, and beareth them from place to place, and changeth place and dwelling, for her nest should not be found. The Weasel pursueth and eateth serpents, and hateth and eateth mice. And their opinion is false that mean that Weasels conceive at mouth, and kitteneth at the ear. And if the Weasel's kittens fall by any hap in chines or in pits, and be hurt or dead, the Weasel healeth them with a certain herb, and reareth them from death to life. And eateth rue, and balmeth herself with the juice thereof, and reseth [rages] then on the cockatrice, and assaileth, and slayeth him without any dread boldly. The Weasel knoweth soon of the cockatrice, and

goeth into his den and slayeth him there. And is a beast that sleepeth much, and waxeth fat with sleep, and hath gall that helpeth much against adders. And so if a man fall into lethargy, the sleeping evil, by venom of an adder, the ashes of a Weasel, tempered with drops of water, dissolveth and destroyeth the strength and might of the sleep. His biting is malicious and venomous.

Bartholomew (*Berthelet*), bk. xviii. § 74.

If Weasels come unto dead men they will pull out their eyes. Hunters hold opinion here in England that, if they meet with a Weasel in the morning, they shall not speed well that day. They have knowledge like mice and rats to run out of houses before their downfall. If the powder of a Weasel be given unto a cock, chickens or pigeons, they shall never be annoyed by Weasels. Likewise if the brain of a Weasel be mingled with a rennet in cheeses, it keepeth them from being touched with mice, or corrupted with age. The powder thereof mixed with water driveth away mice. If one of their tails be cut off, all the residue do forsake the house. The whelp of a Weasel doth cure the venomous bitings of the shrew. The biting of a Weasel is cured by onions and garlic, either applied outward or taken in drink, so that the party drink sweet wine thereon. Sometimes the Weasel biteth some cattle, which presently killeth them, except there be some instant remedy. If they be rubbed with a Weasel's skin, they are instantly healed.

Topsell, "Four-footed Beasts," pp. 562-8.

If the heart of a Weasel be eaten while still palpitating, it makes a man know future events, and if any man eat of its heart with the eyes and tongue of a dog, he will forthwith lose his voice.

Albertus Magnus, "Of the Virtues of Animals."

Wether.

WINTER'S TALE, iv. 3, 33.

V. Sheep.

Whale.

ALL'S WELL THAT ENDS WELL, iv. 3, 249.

When the Whale hungereth sore, he casteth out of his mouth a vapour, that smelleth as the smell of amber; and fish have liking in that smell, and for the odour and smell of that vapour, they go into the Whale's mouth, and be so deceived and eaten. Also in this fish earthy matter hath more mastery than watery, and therefore he is soon great and fat; and so in age for greatness of body, on his ridge [back bone] powder [dust] and earth is gathered, and so digged together that herbs and small trees and bushes grow thereon; so that that great fish seemeth an island. And if shipmen come unwarily thereby, unneath [hardly] they scape without peril,—for he throweth so much water out of his mouth upon the ship, that he overturneth it sometime or drowneth it. And also he is so huge in quantity that when he is taken, all the country is the better for the taking. Also he loveth his whelps with a wonder love, and leadeth them about in the sea long time. And if it happeth that his whelps be let with heaps of gravel and by default of water, he taketh much water in his mouth, and throweth upon them, and delivereth them in that wise out of peril. And he setteth

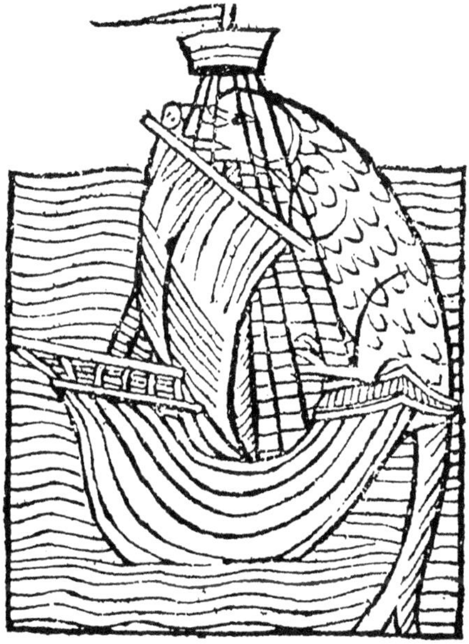

them alway between himself and the sun on the more sicher [safe] side. And when strong tempest ariseth, while his whelps be tender and young, he swalloweth them up into his own womb; and when the tempest is gone and fair weather come, then he casteth them up whole and sound. Also against the Whale fighteth a fish of serpent's kind, and is venomous, as a crocodile; and then other fish come to the Whale's tail, and if the Whale be overcome, the other fish die. And if the venomous fish may not overcome the Whale, then he throweth out of his jaws into the water a fumous smell most stinking; and the Whale

throweth out of his mouth a sweet-smelling smoke, and putteth off the stinking smell, and defendeth and saveth himself and his in that manner wise.

Bartholomew (*Berthelet*), bk. xiii. § 29.

Wheat.

TEMPEST, iv. 1, 61.

WHEN Wheat is gathered, some of the straw is burnt to help and amend the land; and some is kept to fodder of beasts, for it is first meat that is laid tofore beasts. And the kind thereof is cold, that it suffereth not snow that falleth to shed, and is so hot that it compelleth apples for to ripen. Of corrupt dew, that cleaveth to the leaves, cometh corruption in corn, and maketh it as it were red or rusty. And among all manner corn Wheat beareth the price, and to mankind nothing is more friendly, nothing more nourishing. *Bartholomew* (*Berthelet*), bk. xvii. § 65.

OF Wheat is double kind,—one manner kind is red without and sharp at either end, cloven in the side, and is most white within, and heavy in weight, and that manner of Wheat is best; the other manner Wheat is yellow without, and clear and white within, and is light and not easily broken. Grains of Wheat chewed helpeth against the biting of a wode hound, for it draweth out the venom. Also bran of Wheat nourisheth little or else right nought.

Bartholomew (*Berthelet*), bk. xvii. § 168.

Whelp.

i. KING HENRY VI., iv. 7, 35.

WHELPS be the children of hounds. Hounds' Whelps be whelped with sawing teeth though they be full small. And all beasts that have teeth like a saw and departed be gluttons and fight, as the hound, the wolf, the lion, the panther and such other; and all such beasts gender imperfect broods, and the cause is gluttony, for if she should abide until the Whelps were complete and perfect, they should slay the mother with strong sucking, and therefore it needeth that kind be hasty and speedful in such beasts. And authors command to take sucking Whelps wholesomely against venomous bitings, for such Whelps opened and laid hot to the biting of serpents draw out venom. And though they be melancholy beasts of quality and of complexion, yet they be quiver and swift by disposition of numbers, and be glad and merry, and play much, and that is because of their age. *Bartholomew* (*Berthelet*), bk. xviii. § 28.

Wild-duck.

i. KING HENRY IV., ii. 2, 103, iv. 2, 20.

V. **Duck.**

Wild-Goose.

i. KING HENRY IV., ii. 4, 152.

V. **Goose.**

Willow.

OTHELLO, iv. 3, 28, etc.

WILLOW is a pliant tree and a nesh [soft], and according to binding and railing of vines and vine-branches. This tree hath no fruit but only seed or flower. And the seed thereof is of this virtue, that if a man drink of it, he shall get no sons, but only barren daughters. [Query, whether this theory is the reason why the Willow is the badge of the forsaken lover?] Of Willows be perches made and rails for vines. Of the rinds be made bonds and hoops. And [another kind of] Willows be less and more pliant, and therewith men bind wine-pipes and tuns. And of the third kind of Willows be made divers needful things to household, as stools, seats, paniers and cups. Oft in the hollowness thereof lieth venomous worms, as adders and serpents, and therefore it is not sicher [safe] to sleep under the Willow-tree. *Bartholomew* (*Berthelet*), bk. xvii. § 144.

[Gerard has engravings of seven sorts of Willows, including the osier or water-willow.]

Being pilled they are excellent good for the more delicate sort of wicker-ware, and better far than stubborn leather; but principally for leaning-chairs, wherein a man or woman may gently take a nap, sitting at ease, and reposing most sweetly. *Holland's Pliny*, bk. xvi. ch. xxxvii.

[Evelyn ("Silva," p. 101) enumerates twenty-two kinds of "Willow, withy, sallow and osier." He thinks that the ingenious house-wife might make of the willow, cotton, cushions, and pillows of chastity. From the osier Evelyn says are made "baskets, flaskets, hampers, cages, lattices, cradles, and the bodies of coaches and waggons, for which 'tis of excellent use, light, durable and neat, as it may be wrought and covered,—chairs, hurdles, stays, bands, fish-wives, and for all wicker and twiggy works."]

Wine.

The worthiness and praising of Wine might not Bacchus himself describe at the full, though he were alive; for among all liquors and juice of trees, Wine beareth the price, for passing all liquors Wine moderately drunken most comforteth the body, and gladdeth the heart, and healeth and saveth wounds and evils. Wine heateth cold bodies, and cooleth hot bodies, and moisteth dry bodies, and abateth and drieth moist bodies. And in Wine take heed of these things:—of the liquor, of colour, of savour and smell. Of colours of Wine be four manners, white, black, citron [yellow], and red. *Bartholomew Berthelet*), bk. xvii. § 184.

RED Wine that is full red as blood is most strong, and maketh strong drunkenness, and needeth therefore to be right well watered. And such Wine turneth soon to blood because of likeness that it hath with blood in liquor, savour and colour. Also Wine turneth the soul out of cruelness into mildness, out of covetise into largeness, out of pride into meekness, and out of dread into boldness. The drunklew [drunken] man's face is pale, his cheeks hang, his eyes be full of whelks and pimples and of blearedness. The drunklew man's hands tremble and shake, and his tongue is bounden and knit, and his stomach bolketh and giveth up in the morrow-tide some foul and abhominable stinking thing, as it were a pit, wherein some dead carrion lieth, and feeleth and is grieved with sore pricking and aching in his head. And the palate or roof of the mouth waxeth bitter by choler, that is heat; by hot fumosity of kind, the throat is tormented with dryness, burning and thirst; and Wine-drunken men fare as the worms that suck blood, for ever the more the Wine-drunken man drinketh, the more he is athirst. And if Wine be oft taken, anon the body abideth as it were a ship in the sea without stern [rudder] and without lodesman, and as chivalry without prince or duke. Therefore the drunken man favoureth the thing that should not be favoured, and granteth that should not be granted, and praiseth that should not be praised, and maketh of wise men fools, and of good men and well-willed, drunkenness maketh evil men and wicked.

Bartholomew (*Berthelet*), bk. xvii. § 185.

WINE made is made by craft of good spicery and herbs, and such Wines be wholesome and liking, when wholesome spicery and herbs be incorporate therein in due manner; for virtue of spicery keepeth and saveth wines that they be not soon corrupt. *Ibid.*, § 187.

THE juice of grapes is called in English Wine. For certain other juices, as of apples, pomegranates, pears, medlars, or services, or such as otherwise made (for example's sake) of barley and grain, be not at all simply called Wines, but with the name of the thing added whereof they do consist —the Wine which is pressed forth of the pomegranate berries is Wine of pomegranates, out of pears, perry, and that which is compounded of barley is called barley-wine —in English, ale or beer. Wine is a remedy against taking of hemlock or green coriander, the juice of black poppy, wolf's-bane and leopard's-bane, toad-stools, and other cold poisons, and also against the biting of serpents, and stings of venomous beasts, that hurt and kill by cooling.

Gerard's "Herbal," *s.v.* "Vine."

[Among the Wines mentioned by Elizabethan dramatists are: Sack, Claret, Charneco, Canary, Palermo, Sherry, Greek, Spanish, Orleans, French, Vino de Monte, Cyprus, Candy, Graves, Saragossa, Pedro Ximenes (Peter-see-me), Bordeaux, De Clare, Corsican, Malmsey, Hypocras (a compound of Wine and herbs), Lentica, Muscadine, Whippincrust, Rhenish, Lesbian, Drum-wine, White Muscadel, Merry-go-round (slang for wine?), Alicant, Aristippus, Cherally, Madeira, Malaga, Nipitato, Verdea, Fontiniac, Gascon, Nectarella, Deal, Back-rag, Medea, Tunis, and Bastard (white and brown), etc.

Wine was mixed with sugar (*q.v.*), amber (*Beaumont and Fletcher's* "Custom of the Country," *Ben Jonson's* "Magnetic Lady"), with rose-water and sugar, to correct hardness ("London Prodigal"), eggs, carduus (as a cure for obesity, *Beaumont and Fletcher's* "Philaster"), and borage ("Trial of Treasure"). It was also mulled or burnt. From a pint to a gallon was the allowance for a man.

"The Widow's Treasure" gives a test for the purity of Wine, viz., that ripe mulberries or a pear clean pared swim in pure Wine, and sink in watered Wine.]

The vintners sold no other sacks, Muscadels, Malmsies, Bastards, Alicants, nor any other Wines but white and claret, till the 33rd year of King Henry VIII. (1543). All those sweet Wines were sold till that time at the apothecaries' for no other use but for medicines.

John Taylor, "The Old, Old, Very Old Man," (Life of Thomas Parr).

Wolf.

The Wolf hath virtue in his feet, as the lion hath, and so what he treadeth with his feet liveth not. Churls speak of him and say, that a man loseth his voice if the Wolf seëth him first; and certainly, if he know that he is seen first, he loseth his boldness, hardiness and fierceness. The Wolf may not dure with hunger long time, and devoureth much after long fasting. In Ethiopia be Wolves with hair and manes in the neck, and be so speckled, and have so many diverse colours, that they lack no manner colour. Wolves of Africa be nice and little. In Ind is a Wolf that hath three rows of teeth above, and hath feet like a lion, and face as a man, and tail as a scorpion, and his voice is as it were a man's voice, and dreadful as a trumpet; and the beast is swift as an hart, and is right fierce and cruel, and eateth men. Also Wolves, when they flee, bear with them their whelps, and eat *Origanum*, and chew it when they go out of their dens, and to whet and sharp their teeth with. Also the Wolf loveth well to play with a child, if he may take him, and slayeth him afterward, and eateth him at the last. If the Wolf be stoned, he taketh heed of him that throweth the first stone, and if that stone grieveth him, he will slay him, and if it grieveth him not, and he may take him that throweth that stone, he doth him not much harm, but some harm he doth him, as it were in wrath, and leaveth him at last. And the entrails of Wolves be right feeble, and take soon corruption, when they be wounded, and the other deal of the body suffereth many strokes, and hath great strength in the neck and in the head. Also the Wolf desireth kindly to eat fish, and eateth the filth that fishers throw out of their nets; and when he findeth nothing to eat that the fishers leave, then he goeth to their nets, and breaketh and rendeth them. The virtue and strength of Wolves is in the breast and in the claws, and in the mouth, and least in the hinder parts. And the Wolf may not bend his neck backward in no month of the year, but in May alone, when it thundereth; and hath a cruel wariness, so that he taketh no prey of meat nigh to the place where he nourisheth his whelps, but he hunteth in places that be far thence. And when he goeth by night for to take his prey, if it happeth in any wise, that his foot maketh noise treading upon any thing, then he chastiseth that foot with hard biting. His eyes shine by night as lanthorns. And he beareth in his tail a lock of hair that exciteth love,—and doth it away with his teeth when he dreadeth to be taken. The Wolf dreadeth greatly stones, so that if a man take two stones, and smite them together, the Wolf loseth boldness and hardiness, and fleëth away, if the noise of the stones cometh to his hearing. The Wolf eateth earth when he findeth none other prey; and deceiveth sheep more with guile and wrenches [tricks] than with might and strength. He infecteth the wool of the sheep that he slayeth, and his biting maketh the cloth lousy, that is made of that wool. Also a string made of a Wolf's gut, put among harp-strings made of the guts of sheep, destroyeth and corrupteth them, as the eagle's feathers put among culver's pilleth and gnaweth them, if they be there left together long in one place. *Bartholomew* (*Berthelet*), bk. xviii. § 71.

A Wolf doth fear greatly stones; therefore when he is constrained to go by stony places, he treads very demurely or softly; for being hurt with a very little stroke of a stone, it breeds worms, wherof at length he is consumed, or brought to his death.

Lupton, "Notable Things," bk. v. § 3.

The dung of a Wolf, being hidden in a stable or house, where cattle be, especially sheep, it will not only make them leave from eating their meat, but also it will cause them to stir up and down, and to bleat, or to make a noise, and also to quake and tremble, as though their devouring enemy the Wolf were there present. Neither will they cease from doing this, until they feel or perceive that the said dung is taken away (*Mizaldus*).

Ibid., bk. vi. § 80.

A Wolf first seeing a man doth lift up his voice, and as a victor doth despise him; but if he perceive that the man hath espied him first, he lays away his fierceness, and cannot run. *Ibid.*, bk. viii. § 2.

If the head of a Wolf be hanged in a dove-house, neither cats, weasels, nor any thing that will hurt pigeons will enter therein. *Ibid.*, bk. ii. § 78.

Bells covered with the skin of a Wolf do drown the sound of other bells that are covered with the skin of a lamb. *Ibid.*, § 92.

Whosoever anoints his feet or hands with the grease of a Wolf, he shall not be hurt with any cold of his hands or feet so anointed (*Mizaldus*). *Ibid.*, bk. iii. § 44.

MEN in ancient time did fasten upon the gates of their towns the heads of Wolves,—thereby to put away witchery, sorcery or enchantment. Which many hunters observes or do at this day, but to what use they know not.

Lupton, "Notable Things," bk. iii. § 5.

IF the tail of a Wolf be buried or put in the ground of any town or village, no Wolf will enter in that town or village. *Ibid.*, § 20.

IF one make a little rope of the guts of a Wolf, and then bury the same under the sand or earth, there will neither horse nor sheep go that way, though you beat them with a staff. *Ibid.*, bk. ix. § 98.

THEY say there is antipathy between Wolves, and squill-roots. *Camden's* "Britannia," col. 907 (ed. 1722).

THE Wolves of Scanzia, by reason of extremity of cold in those parts, are blind and lose their eyes. The golden Wolf is exceeding strong, especially being able with his mouth and teeth to bite asunder not only stones but brass and iron. In the Dog-days he hideth himself in some pit or gaping of the earth, until that sunny heat be abated. There be some have thought that dogs and Wolves are one kind: namely that vulgar dogs are tame Wolves, and ravening Wolves are wild dogs. But Scaliger hath learnedly confuted this opinion, shewing that they are two distinct kinds, not joined together in nature, nor in any natural action. The brains of a Wolf do decrease and increase with the moon. The neck of a Wolf is short which argueth a treacherous nature. If the heart of a Wolf be kept dry, it rendereth a most pleasant or sweet-smelling savour. They will go into the water two by two, every one hanging upon another's tail, which they take in their mouths. Wolves do also eat a kind of earth called *Argilla*, which they do not for hunger, but to make their bellies weigh heavy, to the intent, that when they set upon an horse, an ox, a hart, an elk, or some such strong beast, they may hang fast at their throats till they have pulled them down, for by virtue of that tenacious earth, their teeth are sharpened, and the weight of their bodies increased; but when they have killed the beast before they touch any part of his flesh, they empty their bellies of the earth as unprofitable food. If there be many of them, in hunting together they equally divide the prey among them all. One saw a Wolf in a wood take in his mouth a piece of timber of some 30 or 40 pound weight, and with that he did practise to leap over the trunk of a tree that lay upon the earth; at length when he perceived his own ability and dexterity in leaping with that weight in his mouth, he did there make his cave and lodged behind that tree. At last it fortuned there came a wild sow to seek for meat along by that tree, with divers of her pigs following her, of different age, some a year old, some half a year, and some less. When he saw them near him, he suddenly set upon one of them, which he conjectured was about the weight of wood which he carried in his mouth, and when he had taken him, whilst the old sow came to deliver her pig at his first crying, he suddenly leaped over the tree with the pig in his mouth, and the poor sow could not leap after him, and yet might stand and see the Wolf to eat the pig which he had taken from her. When they will deceive goats, they come unto them with the green leaves and small boughs of osiers in their mouths, wherewithal they know goats are delighted, that so they may draw them therewith, as to a bait to devour them. Their manner is when they fall upon a goat or a hog, not to kill them, but to lead them by the ear with all the speed they can drive them to their fellow-Wolves; and if the beast be stubborn, and will not run with him, then he beateth his hinder-parts with his tail, holding his ear fast in his mouth. But if it be a swine that is so gotten, then they lead him to the waters, and there kill him, for if they eat him not out of cold water, their teeth doth burn with an untolerable heat. If a horse tread upon the footsteps of a Wolf which is under a horse-man or rider, he breaketh in pieces, or else standeth amazed. If a Wolf treadeth in the foot-steps of a horse which draweth a waggon, he cleaveth fast in the road as if he were frozen. The Wolf is afraid of a sea-crab or shrimp. If a man anoint himself with the fat or suet taken out of the reins of a lion, it will drive away from him all kinds of Wolves. The Ravens are in perpetual enmity with Wolves, and the antipathy of their natures is so violent that if a raven eat of the carcase of a beast which the Wolf hath killed, or formerly tasted of, she presently dieth. The sea-onion of all other things is hateful to a Wolf, and by treading on it his leg falleth into a cramp; the Wolf is an enemy to the fox and turtle, and in their absence from their nests, they leave this onion in the mouth thereof, as a sure guard to keep their young ones from the Wolf. A she-Wolf the first year littereth one whelp, the second year two, the third year three, and so observeth the same proportion unto nine, after which she groweth barren; and when she bringeth her young ones to the water, if any of them lap water like a dog, him she rejecteth as unworthy of her parentage, but those which suck their water like a swine, or bite at it like a bear, them she taketh to her, and nourisheth very carefully. Of dogs and Wolves cometh the panther, of the hyæna and the Wolf come the *Thoes*, and the hyæna itself seemeth to be compounded of a Wolf and a fox. The skin of a Wolf being tasted of those which are bit of a mad or ravenous dog doth preserve them from the fear or hazard of falling into water. If any labouring or travelling man doth wear the skin of a Wolf about his feet, his shoes shall never pain or trouble him. He which doth eat the skin of a Wolf well tempered and sodden will keep him from all evil dreams, and cause him to take his rest quietly. The teeth of a Wolf being rubbed upon the gums of young infants doth open them whereby the teeth may the easier come forth.

Topsell, "Four-footed Beasts," pp. 568-84.

If any man bind the right eye of a Wolf on his right sleeve, neither men nor dogs can hurt him.

Albertus Magnus, "Of the Virtues of Animals."

Woodbine.

MUCH ADO ABOUT NOTHING, iii. 1, 30.

[Honeysuckle; but in "Midsummer Night's Dream," iv. 1, 47, Nares thinks it is used for bindweed or convolvulus.]

Woodcock.

O this woodcock! what an ass it is!

TAMING OF THE SHREW, i. 2.

THE Woodcock strives to hide his long bill, and then thinks nobody sees him.

Dean Swift, "Polite Conversation," dialogue i.

[Jonston, "Natural History of Birds," ch. iv. (1657), says that it was popularly supposed to be without gall.

Woodcocks were baked (*Dekker's* "Raven's Almanac") and made into pies (*Webster's* "Westward Ho!"), and served on buttered toast (*Ben Jonson's* "Epicure," *Middleton's* "New Way to pay Old Debts").]

Worm.

A Worm is a beast that oft engendereth of flesh and of herbs, and gendereth oft of gall, and sometime of corruption of humours, and sometime of meddling of male and female, and sometime of eggs, as it well appeareth of scorpions and of tortoises and newts. And worms come out of their dens in springing - time. Of Worms be many manner diverse kinds, for some be water-worms, and some be land-worms; and of those some be in herbs and in worts, as malshrags [caterpillars] and other such; and some be in trees, as tree-worms; and some in clothes, as moths; and some in flesh as maggots, that breed of corrupt and rotted moisture in flesh; and some in beasts within and without, as long Worms in children's wombs, and lice and nits in heads, and all such Worms breed and gender of corrupt humours in bodies of beasts within and without. And there be other Worms of the earth, which be long and round, nesh [soft] and smooth, as angle - twitches, and moles do hunt them under earth, and with angle-twitches fish is taken in waters, when fish-hooks be baited with such Worms instead of bait. And such Worms help against cramp, and against shrinking of sinews, and also against biting of serpents and against smiting of scorpions. And among Worms, some be foot-less, as adders and serpents, and some nave six feet, and some be full evil and malicious, and enemies to mankind, as serpents and other venomous Worms; and some Worms be round of body, and hath no sinews nor bones, great nor small, neither gristles, neither blood;—and all such dieth if they be anointed with oil, and do quicken again in vinegar. And some Worms gender and be gendered, and some be gendered and gender not, as the Salamander.

Bartholomew (*Berthelet*), bk. xviii. § 115.

Of earth-worms some are bred only in the earth, and others among plants, and in the bodies of living creatures. Worms are found to be very venomous in the kingdom of Mogor, and the inhabitants there do stand in so great fear of them, that they be destroyed and slain by them when they travel any journey; and therefore there they ordinarily use to carry besoms with them to sweep the plain ways for fear of further hurt.

Topsell, "History of Serpents," pp. 811-13.

If you stamp earth-worms, and then strain them through cloth, and then put to the same as much of the oil of radish-roots, and between the beating or framing of swords, knives, or daggers, when they be hot, you do quench them twice or thrice therein, the same shall cut iron after, as though it were lead. *Lupton*, "Notable Things," bk. ii. § 43.

If Worms gnaw upon or hurt the mouth of the stomach, put honey-combs into the mouth fasting, and hold them there, and the Worms will draw into the honey, and so void by the mouth. It hath been proved.

Ibid., bk. iv. § 56.

Worms and other venomous beasts are driven away from any place with the smoke or fume of other beasts of the same kind. *Ibid.*, bk. x. § 67.

V. Vermin.

Wormwood.

Hamlet, iii. 2, 191.

Wormwood is a full sharp herb, and is gathered in the end of springing-time, and dried in shadow. And syrup made of Wormwood exciteth appetite and withstandeth drunkenness. Wormwood with powder of cummin and honey doth away moles and speckles, and ache that cometh of smiting. And Wormwood keepeth and saveth books and clothes from fretting of mice and of worms, if it be laid therewith in chests or coffers. And helpeth against biting of weasels and of dragons, and healeth it, if it be drunk. Wormwood exciteth sleep, if it be laid unwittingly under the head; and maketh black hair, if the hair be anointed with ointment made of the juice thereof and oil of rose. *Bartholomew* (*Berthelet*), bk. xvii. § 12.

Let a man or woman use to drink Wormwood, they shall not be sea-sick. If writing-ink be tempered with the infusion of Wormwood, it preserveth letters and books written therewith from being gnawn by mice.

Holland's Pliny, bk. xxvii. ch. vii.

WHILE Wormwood hath seed, get a handful or twain,
To save against March, to make flea to refrain,
Where chamber is sweeped, and Wormwood is strown,
No flea for his life dare abide to be known.
What savour is better, if physic be true,
For places infected, than Wormwood and rue?
It is as a comfort for heart, and the brain,
And therefore to have it, it is not in vain.

Tusser, "July's Husbandry," st. 10 and 11.

Worts.

MERRY WIVES OF WINDSOR, i. 1, 124.

How to make long Worts:—Take a good quantity of coleworts, and seethe them in water whole a good while, then take the fattest of powdered beef-broth, and put it to the Worts, and let them seethe a good while after; then put them in a platter, and lay your powdered beef upon it.

"The Good Huswife's Handmaid," p. 8, b.

V. Cabbage.

Wren.

MACBETH, iv. 2, 9.

IT is much to be marvelled at the little bird called a Wren, being fastened to a little stick of hazel newly gathered, doth turn about and roast himself.

Lupton, "Notable Things," bk. vii. § 57.

Yew.

MACBETH, iv. 1, 27.

A YEW-TREE is a tree with venom and poison, and is a strong tree and an high, with great boughs pliant and long; such trees are burnt and bows made thereof. The shadow thereof is grievous, and slayeth such as sleep thereunder. *Bartholomew* (*Berthelet*), bk. xvii. § 161.

THE birds that eat the red berries either die or cast their feathers. *Batman's* addition to *Bartholomew*, *loc. cit.*

[Gerard denies that the Yew-berries are poisonous, and that its shadow is dangerous; but to this day the Yew is held to be poisonous among country-folk.]

GLOSSARY.

(*The figures refer to one or more of the pages on which the word occurs.*)

Abhominable, 345. See *Love's Labour's Lost*, v. 1, 26
Able, suitable, 272
Abrood, brooding on the nest, 96
Accord, reconcile, 53
According, suitable, 157
Againward, on the other hand, *vice versâ*, 332
Alloren-tree, alder, 246
All to, quite, altogether, 75
Almery, cupboard, or recess, 30
Angle-twitch, earth-worm, 352
Aposthumation, imposthume, abscess, 211
Appair, impair, injure, 102
Arear, rise, 45
Arred, increased, 108
Attercop, spider, 33
Away with, endure, bear, 19

Barnacle, bit, 20
Batable, luxuriant, 136
Batableness, luxuriance, fertility, 134
Bauer, boor, 245
Beclip, embrace, 3, 30
Behindforth, from behind, 178
Bespring, sprinkle, 33
Bill, peck, 292
Bird, young bird, 97; young fish, 111
Blazing, bright, 6
Blowing, maggot, 158
Boisterous, coarse, rough, 169
Bolk, belch, 345
Botch, sore, 86
Bray, pound, 115
Brent, burnt, 102
Bugle, buffalo, 45
Burgeon, bud, sprout, 294
Bush, push, butt, 255
Busy, to be busy, strive, 17
But, unless, 130

Calvered, pickled, 297
Cap, a large-sized paper, 106
Castein, chestnut, 62
Casting, vomiting, 87
Cease, make to cease, 87
Ceruse, white lead, 20
Charge, burden, 97 (here "extra growth")
Chauffe, warm, 222
Cheese-lope, rennet, 290
Cherry-cracker, spotted fly-catcher (?), 37
Chine, crevice, hollow, 20
Chirking, chirping, 7
Citrine, yellow, 191
Cliff, cleft, 281
Coarcting, drawing together, 239
Codware, leguminous plant, 28
Colewort, cabbage, and its kinds, 266
Commonty, commonwealth, sometimes common people, 96
Continual, continuous, 31
Convenable, suitable, convenient, 238
Cop, crest (?), 216. The **cop** is the cone of corn above the rim of a bushel or other measure
Copped, crested, 173
Corrump, corrupt, 97
Cotiniate, confection, 254
Covetise, covetousness, 11
Cratch, manger, 103
Cross or **Gang Week**, Rogation Week, 35. (See Brand's "Popular Antiquities," vol. i., p. 172)
Culver, dove, 91

Damask, Damascus, 82
Deal, part, 60
Deceivable, deceitful
Defoil, damage, 102
Defy, digest, 150
Disparkled, or **Disparpled**, dispersed, scattered, 96, 327

Distingue, distinguish, mark, 220, 305
Drasts, dregs, 329
Drunklew, drunken, 345
Dunbird, female pochard, 37
Dure, last, endure, 285

Ean or **Yean**, bring forth, 285
Ellern, elder-tree, 100

Fail, to be hidden, 270
Farcing, stuffing, 156
Fast, close-grained, 126
Fear, frighten, 20
Fen, mud, 75
Festue (fescue), a pointed piece of straw, 153
Fime, dung, 83
Flasket, bottle, 158
For, because, 17; so that, 195
Fordo, hinder, prevent, 169
Fore, trace, going, 183
Forthright, straight ahead, 17
Fresh-lap, dew-lap, 227
Frog-stool, toad-stool, 216
Frore, frozen, 270
Frote, rub, 19
Fumosity, smoke, fume, humour, 301

Garled, variegated, 170
Gin, trap, 121
Glead, kite, hawk, 172
Glimy, or **Glymy**, viscous, gluey, 215
Glires (Lat.), dormice, 90
Gramuel, crayfish, 275
Graving, engraving, 117
Grede, scream, 77
Gripe, griffin, 136
Grudge, to be grieved, 20, 173
Grutching, growling, 305

Hamborough, horse-collar, 278
Hard, harden, 303
Haught, grown, 235
Haver, oats, 218
Haver-cake, oat-cake, 218
Hele, close, conceal, 20
Herb-grace, rue, 266
Hobby, a small hawk, 147
Holp, helped, 102
Howlet, owlet, 302
Hull, husk, 218

Imposthume, abscess, 210
In-wit, mind, 63

Jack, a hawk, 147
Jannock, a large loaf of oaten bread, 218
Jump, swindle, 262
Jumper, swindler, 262

Keep, take care, 158
Kind (*adj.*), natural, genuine, 7
Kind (*subst.*), Nature, 9
Kindly (*adv.*), by nature, 19
Knot, flower-bed, 126

Lacture, lettuce, 156
Lanner, **Lanneret**, hawks, 147
Latten, v. p. 174, a kind of brass
Laund, lawn, 102
Let, hinder, 105, 212, etc.
Ley, meadow, 20
Lodesman, steersman, 345
Looch, lozenge, "medicine to be licked," 200

Make, mate, 17
Malshrag, caterpillar, 48
Mammet, or **Maumet**, idol, image (derived from "Mahomet"), 218
Manner, kind of, 9
Marais, marsh, 122
Margarite, pearl, 105
Marlion, a merlin (a kind of hawk), 147
Mean, think, opine, 50
Meddle, mix, 28
Merigall, gall, chafe, 314
More, root, 291
More, greater, 305
Morrow, morning, 238
Mother, scum, 206
Musket, male sparrow-hawk, 147

Namely, especially, 33
Natheless, nevertheless, 85
Navew, turnip, 319
Ne, nor, 17
Neat, cattle, 227
Neep, turnip, 319
Neeze, sneeze, 266
Nep, mint, cat's-mint, 115
Nesh, soft, 6
Nesh, to make soft, 130
Neshness, softness, 175
Nich, a bundle, 115
Nigh, to approach, 327
Nippitate, a wine, 300
Nock, notch, 227
Noy, annoy, 3
Noyful, annoying, grievous, 128

Oliet, or **Olife**, perhaps olive or oyster-catcher, 37
Out-take, except, 9
Overset, attacked, 11

Paddock, toad, 230
Paigle, oxlip, cowslip, 228
Parbraking, vomiting, 87
Pass, surpass, 105
Pavis, shield, fence, 304
Pawper, purple heron (?), 37
Peasen, spawn, 111
Pill, peel, 106
Pine-apple, fir-cone, 246
Plant, sole, 282

Plight, twist, 324
Podagre, gout, 159
Pointel, point, style, 338
Pompion, or **Pumpion**, a pumpkin, melon, or vegetable-marrow, 251
Posthume, imposthume, abscess, 210
Pounce, claw of a hawk, 107
Powder, dust, 5
Prance, to rebel (?), 173. Minsheu has: "the prancing of a horse, *tumultuose flare et strepere*" (ed. 1629). Florio uses "prance" in the modern sense (ed. 1659)
Pregnable, able to take, or penetrate, 131
Prick, prickle, spine, small bone, 139
Pricket, a wax candle of a certain size, 266; a buck of the second year, 251
Pule away, pine away, 132
Pullen, poultry, 432
Puttock, a kite, glead, 172

Quiver, quick, nimble, 45

Race, or **Raze** (of ginger), a root, 126
Rank, rage, 306
Raphan, radish, 219
Rear, raise, 129
Rebound, make to rebound, 96
Refraining, restraint, 21
Rese (*verb*), to rage, rush angrily, 17
Rese (*subst.*), rage, rush, 20
Rib, dressed with a "rib"—*i.e.*, an instrument used in dressing flax, 115
Ridge, back-bone, 89
Rivel, wrinkle, 109
Rivelled, wrinkled, divided (as a wasp's thorax and abdomen), 8
Rivelly, wrinkled, 241
Rivet, roe, 114
Rob, dried juice of herbs (jam ?), 126
Rosere, a rose-tree, 303
Rothern, cattle, 32

Sad, heavy, 126
Savourly, daintily, with pleasure, 281
Sawing, **Sawy**, serrated, 31
Seld, seldom, 129
Senvey, mustard, 212
Several, separate, 222
Shed, melt, flow, 299
Shern-birds, hornets, 32
Shingle, "a cloven board, a plank" (Cooper's "Thesaurus"), 198
Shittle, inconstant, weak, 193
Sicher, safe, 41
Sicherly, safely, 265
Sizing (themselves), limiting themselves to an allowance, 309. So *King Lear*, ii. 4, 178
Slipper, slippery, 4
Smape, crush, 124
Smaragdus, or **Smaragd**, emerald, 137
Some-deal, partly, 289
Sort, quantity, 57
Sparhawk, sparrow-hawk, 147
Spouse-breach, adultery, 177
Spouse-head, wedlock, 183
Spring, sprinkle, 4, 16
Springing-time, spring, 30
Sprung, sprinkled, 115
Spurling, or **Sparling**, smelt, 297
Squinanth, "camel's hay," a kind of rush, 267
Startle, leap suddenly, 89
Stie, climb, 31
Stockfish, dried fish. See 297
Stint, check, stop, cease, stay, 5
Stony, astonish, 199
Stover, fodder, 299
Studdery, stud, stable, 162
Sucket, sweetmeat, 335
Sunstead, solstice, 172
Sweven, dream, 3

Tapping, perhaps misprint for "topping," evidently the horn tips to the bow, 227
Tassel, **Tiercel**, **Tassel-gentle**, or **Tiercel-gentle**, a male hawk, 302
Tawing, beating, 297. So "taw" or "tawes," a (leather) whip
Tharf-cake, or **thard-cake**, a thin, round, unleavened cake of rye or barley, baked hard, 218
Theriaca, treacle, 100
Thirl, pierce, penetrate, 167
Tiercel, a hawk, 147. *V.* **Tassel**
Tiercel-gentle, a hawk, 302. *V.* **Tassel**
Timber, timbrel, tabor, drum, 33
Tivit, perhaps a misprint for "twite," 37
Tofore, before, 299
Tohaul, drag, 157
Toll, draw, 231
Trace, perhaps should be "tract," otherwise the word means "place where it may be traced," 141
Treacle, mithridate, antidote to poison, 9
Troubly, troubled, 175

Underset, support, 77
Unfastness, loose grain (of wood), 303
Union, pearl, 105
Unneath, or **Uneath**, hardly, scarcely, 20
Unsadness, lightness, 303
Unwitting, without the knowledge of, 44
Urchin, hedgehog, 149

Ventuous, windy, 157
Very, true, genuine, 84
Void, avoid, 100

Wagging, loose, 261
Wallowing feet aside, the uneven movement produced by the difference in the length of the fox's legs—"wallow" being a wave, 119

Wambling, rumbling, borborygmi, 203
Want, mole, 204
Wash-tail, wag-tail (?) or dish-washer, 37
Wem, blemish, 271
Wend, turn, 115
Weybread, plantain, 247
Whelk, pustule, pimple, 57. So *King Henry V.*, iii. 6, 108
Whur, whir, 128
Womb, belly, bowels, 30
Wonderly, wonderfully, 8
Wood, or **Wode**, mad, 86
Woodgnaw, great spotted woodpecker (?), 37
Woodness, madness, 3
Woodspike, green woodpecker, 37
Worship, make worshipful, 27
Wreak (*subst.* and *verb*), revenge, 17
Wreakful, revengeful
Wrench, trick, 348
Writhen, twisted, 32
Wroting, rooting, 90

Yean, or **Ean**, bear, bring forth, 129
Yexing, hiccough, 203

THE END.

Elliot Stock, 62, *Paternoster Row, London*, *E.C.*

REST PRAY SLEEP

Printed in Dunstable, United Kingdom

91219956R00208